Everyday
Mathematics®

The University of Chicago School Mathematics Project

STUDENT MATH JOURNAL
VOLUME 1

Mc
Graw
Hill
Education

Bothell, WA • Chicago, IL • Columbus, OH • New York, NY

The University of Chicago School Mathematics Project

Max Bell, Director, *Everyday Mathematics* First Edition; James McBride, Director, *Everyday Mathematics* Second Edition; Andy Isaacs, Director, *Everyday Mathematics* Third, CCSS, and Fourth Editions; Amy Dillard, Associate Director, *Everyday Mathematics* Third Edition; Rachel Malpass McCall, Associate Director, *Everyday Mathematics* CCSS and Fourth Editions; Mary Ellen Dairyko, Associate Director, *Everyday Mathematics* Fourth Edition

Authors
Jean Bell, Max Bell, John Bretzlauf, Amy Dillard, Robert Hartfield, Andy Isaacs, James McBride, Rachel Malpass McCall, Kathleen Pitvorec, Peter Saecker

Fourth Edition Grade 1 Team Leader
Rachel Malpass McCall

Writers
Meg Schleppenbach Bates, Kate Berlin, Sarah R. Burns, Gina Garza-Kling, Linda M. Sims

Open Response Team
Catherine R. Kelso, Leader; Kathryn M. Rich

Differentiation Team
Ava Belisle-Chatterjee, Leader; Anne Sommers

Digital Development Team
Carla Agard-Strickland, Leader; John Benson, Gregory Berns-Leone, Juan Camilo Acevedo

Virtual Learning Community
Meg Schleppenbach Bates, Cheryl G. Moran, Margaret Sharkey

Technical Art
Diana Barrie, Senior Artist; Cherry Inthalangsy

UCSMP Editorial
Lila K.S. Goldstein, Senior Editor; Rachel Jacobs, Kristen Pasmore, Delna Weil

Field Test Coordination
Denise A. Porter

Field Test Teachers
Mary Alice Acton, Katrina Brown, Pamela A. Chambers, Erica Emmendorfer, Lara Galicia, Heather A. Hall, Jeewon Kim, Nicole M. Kirby, Vicky Kudwa, Stephanie Merkle, Sarah Orlowski, Jenny Pfeiffer, LeAnita Randolph, Jan Rodgers, Mindy Smith, Kellie Washington

Contributors
William B. Baker, John Benson, Jeanine O'Nan Brownell, Andrea Cocke, Jeanne Mills DiDomenico, Rossita Fernando, James Flanders, Lila K.S. Goldstein, Allison M. Greer, Brooke A. North, Penny Williams

Center for Elementary Mathematics and Science Education Administration
Martin Gartzman, Executive Director; Meri B. Forhan, Jose J. Fragoso, Jr., Regina Littleton, Laurie K. Thrasher

External Reviewers

The *Everyday Mathematics* authors gratefully acknowledge the work of the many scholars and teachers who reviewed plans for this edition. All decisions regarding the content and pedagogy of *Everyday Mathematics* were made by the authors and do not necessarily reflect the views of those listed below.

Elizabeth Babcock, California Academy of Sciences; Arthur J. Baroody, University of Illinois at Urbana-Champaign and University of Denver; Dawn Berk, University of Delaware; Diane J. Briars, Pittsburgh, Pennsylvania; Kathryn B. Chval, University of Missouri–Columbia; Kathleen Cramer, University of Minnesota; Ethan Danahy, Tufts University; Tom de Boor, Grunwald Associates; Louis V. DiBello, University of Illinois at Chicago; Corey Drake, Michigan State University; David Foster, Silicon Valley Mathematics Initiative; Funda Gönülateş, Michigan State University; M. Kathleen Heid, Pennsylvania State University; Natalie Jakucyn, Glenbrook South High School, Glenview, IL; Richard G. Kron, University of Chicago; Richard Lehrer, Vanderbilt University; Susan C. Levine, University of Chicago; Lorraine M. Males, University of Nebraska-Lincoln; Dr. George Mehler, Temple University and Central Bucks School District, Pennsylvania; Kenny Huy Nguyen, North Carolina State University; Mark Oreglia, University of Chicago; Sandra Overcash, Virginia Beach City Public Schools, Virginia; Raedy M. Ping, University of Chicago; Kevin L. Polk, Aveniros LLC; Sarah R. Powell, University of Texas at Austin; Janine T. Remillard, University of Pennsylvania; John P. Smith III, Michigan State University; Mary Kay Stein, University of Pittsburgh; Dale Truding, Arlington Heights District 25, Arlington Heights, Illinois; Judith S. Zawojewski, Illinois Institute of Technology

Note

Too many people have contributed to earlier editions of *Everyday Mathematics* to be listed here. Title and copyright pages for earlier editions can be found at http://everydaymath.uchicago.edu/about/ucsmp-cemse/.

www.everydaymath.com

Send all inquiries to:
McGraw-Hill Education
STEM Learning Solutions Center
8787 Orion Place
Columbus, OH 43240

ISBN: 978-0-02-143078-9
MHID: 0-02-143078-0

Printed in the United States of America.

1 2 3 4 5 6 7 8 9 RMN 19 18 17 16 15 14

Contents

Unit 4

Unit 5

Activity Sheets

How many counters do you have?

I have _____ counters.

Tell or show your partner how you counted.

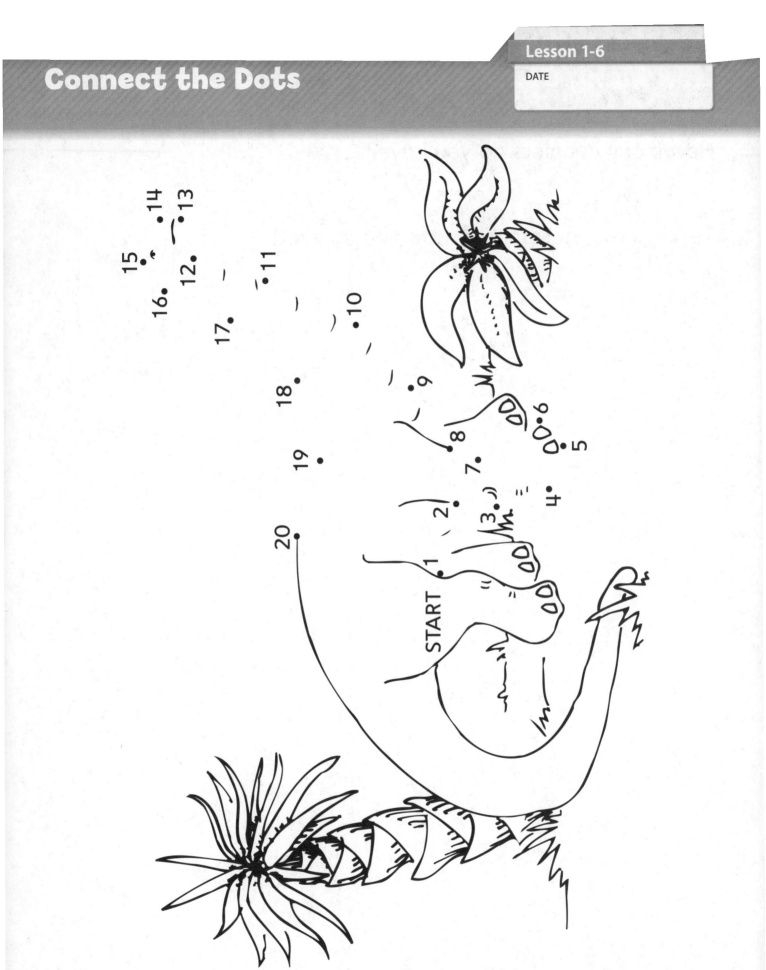

Solve.

Show your work.

Use drawings, numbers, or words.

1 On Monday, Ellie found 2 pennies.

On Tuesday, Ellie found 3 pennies.

How many pennies did Ellie find in all?

_____ pennies

2 On Monday, Tommy found 3 pennies.

On Tuesday, Tommy found 2 pennies.

How many pennies did Tommy find in all?

_____ pennies

Left Handed or Right Handed?

 1 Do you write with your left hand or your right hand?

_____

2 Do you think more children in your class write with their left hand or their right hand?

_____

3 Collect data on the tally chart below to find out.

	Number of Children
Left handed	
Right handed	

4 Are more children in your class left handed or right handed?

_____

How many more? _____

5 How did you figure out how many more?

_____

_____

_____

Math Boxes

1 Count up by 1s.

7,   8,   9,

_____, _____, _____,

_____, _____, _____,

_____, _____

2 How many windows are in your classroom?

_____ windows

3 How many tally marks?

~~||||~~ |

Choose the best answer.

◯ 5
◯ 11
◯ 6
◯ 1

4 Amy had 4 pennies.

Ⓟ Ⓟ Ⓟ Ⓟ

John gave Amy 1 penny.

Ⓟ

How many pennies does Amy have now?

_____ pennies

Picking 10 Apples Record Sheet

You pick 10 apples.

Some of the apples are red.

Some of the apples are green.

How many of each color apple can you pick?

Red Apples	Green Apples

Math Boxes

1 Use your number grid.

Start at 12.

Count up 5.

You end at _____.

2 Josie wins 3 games. Then she wins 4 more games.

How many games did Josie win in all?

_____ games

3

The Pets We Own	
Pet	Tallies
Cat	~~HHH~~ ~~HHH~~
Dog	~~HHH~~ ///
Other	~~HHH~~ /

How many cats? _____ cats

How many dogs? _____ dogs

4 Count by 1s.

8   9   _____   _____   _____   _____   _____   _____

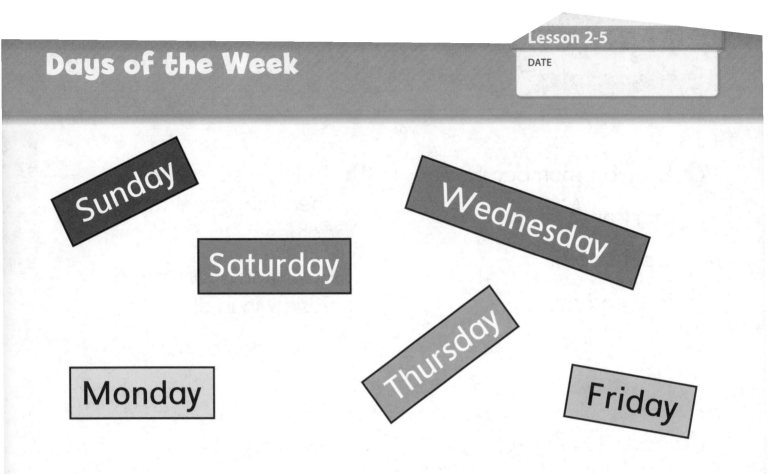

Are any days missing?

_____

Which days are missing?

_____

Tell your partner how you know.

Math Boxes

1 Count back by 1s.

18,   17,   16,

_____, _____, _____,

_____, _____, _____,

_____, _____

2 How many chairs are in your classroom?

_____ chairs

3 How many tally marks?

~~HHT~~ ~~HHT~~

_____ tally marks

~~HHT~~ ~~HHT~~ ///

_____ tally marks

4 Edna drew 4 triangles.

She drew 3 circles.

How many shapes did Edna draw in all?

_____ shapes

nine   9

Math Boxes

1 Use your number grid.
Start at 8.
Count up 7.
You end at _____.
Choose the best answer.

- ⬭ 16
- ⬭ 15
- ⬭ 14
- ⬭ 1

2 Eli has 10 crayons.
Sammy takes 2 crayons
from Eli.
How many crayons does
Eli have now?

_____ crayons

3 Fill in the table.

Children in My Class		
Children	**Tallies**	**Total**
Boys		
Girls		

4 Count by 1s.

22   23   ____ ____ ____ ____ ____ ____

10   ten

Math Boxes

① Draw one way you and your partner could share 5 pencils.

② How many dots are there? Can you tell without counting?

_____ dots

③ Use your number grid.
Start at 10.
Count up 7.

You end at _____.

④ Mel has 2 gold stars and 1 silver star. How many stars does Mel have?

_____ stars

Mary has 1 gold star and 2 silver stars. How many stars does Mary have?

_____ stars

Math Boxes
Preview for Unit 3

Math Boxes

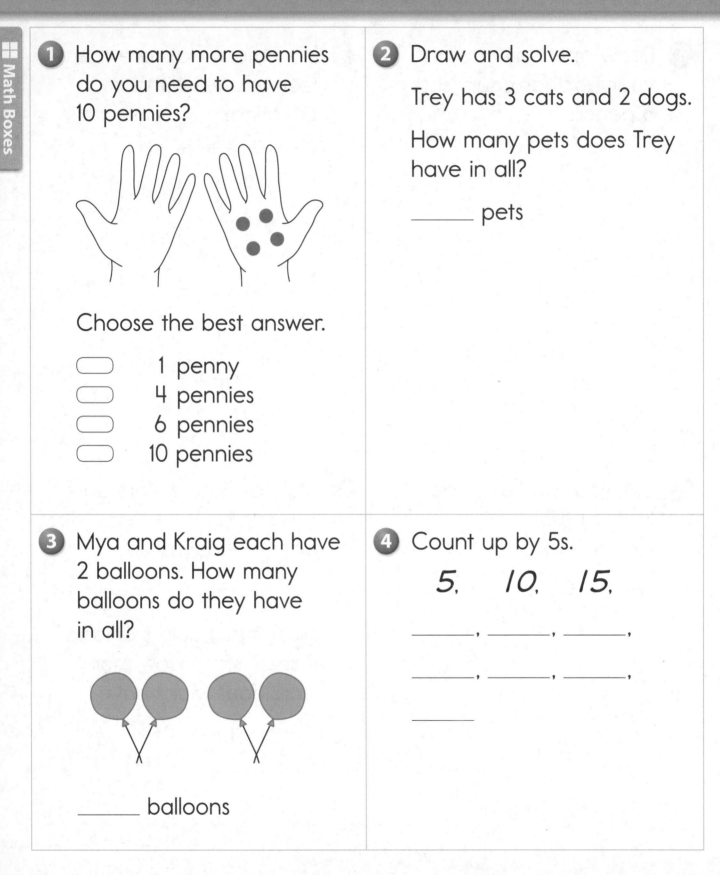

1 How many more pennies do you need to have 10 pennies?

Choose the best answer.

- ⬭ 1 penny
- ⬭ 4 pennies
- ⬭ 6 pennies
- ⬭ 10 pennies

2 Draw and solve.

Trey has 3 cats and 2 dogs.

How many pets does Trey have in all?

_____ pets

3 Mya and Kraig each have 2 balloons. How many balloons do they have in all?

_____ balloons

4 Count up by 5s.

5, 10, 15,

_____, _____, _____,

_____, _____, _____,

_____

Math Boxes

 Math Boxes

1 Caleb had 0 pet fish.
Then he got 3 pet fish.
How many pet fish does
Caleb have now?

_____ fish

2 Trudy is playing *Bunny Hop*.
She starts at 6 and then
moves to 8.
What number did she roll?

_____

3 Count to complete this part of the number grid.

1								9	
		13		15			18		
	22				26				30
		34			37				

4 Use your number line.
Start at 9.
Count back to 4.
How many did you count back? _____

Math Boxes

1 Which numbers add to 5?

Choose the best answer.

- ⬭ 5 and 5
- ⬭ 5 and 1
- ⬭ 2 and 3
- ⬭ 3 and 3

2 How many dots in all?

_____ dots

3 Use your number grid.
Start at 11.
Count up 9.

You end at _____.

4 Sanjay played 5 songs
in the morning.
He played 3 songs
in the afternoon.
How many songs did
Sanjay play in all?

_____ songs

Maria played 3 songs
in the morning.
She played 5 songs
in the afternoon.
How many songs did
Maria play in all?

_____ songs

Writing and Solving Number Models

Write a number model for each number story.
Use a box to show which number is unknown. Then solve.
Use words or pictures to tell how you solved the problem.

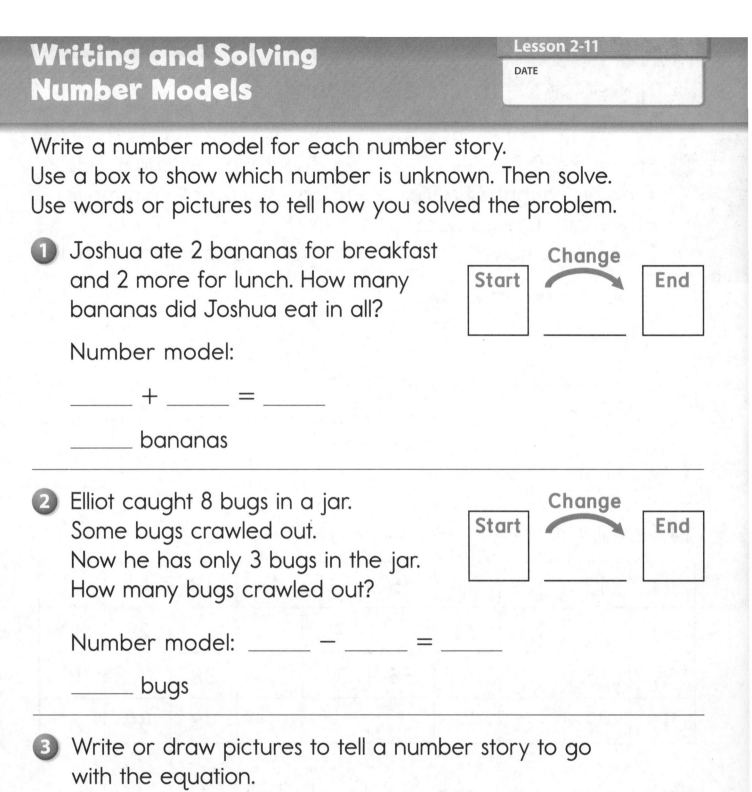

1 Joshua ate 2 bananas for breakfast and 2 more for lunch. How many bananas did Joshua eat in all?

Start   Change   End

Number model:

_____ + _____ = _____

_____ bananas

2 Elliot caught 8 bugs in a jar.
Some bugs crawled out.
Now he has only 3 bugs in the jar.
How many bugs crawled out?

Start   Change   End

Number model: _____ − _____ = _____

_____ bugs

3 Write or draw pictures to tell a number story to go with the equation.

$9 - 7 = 2$

Math Boxes

1 Brody had 0 fireflies. Then he caught 7 fireflies. How many fireflies does Brody have now?

Choose the best answer.

0 + 7 = _____

- ⬭ 0
- ⬭ 6
- ⬭ 7
- ⬭ 17

2 Zoe is playing *Bunny Hop*. She starts at 4 and moves to 8. What number did she roll?

_____

3 Count to complete this part of the number grid.

11	12			15		17	18		
21		23	24		26			29	30
31			35	36		38			40
41	42		44				48	49	

4 Use your number grid. Find the difference.

13 − 4 = _____

1. Make a sum of 10 pennies.

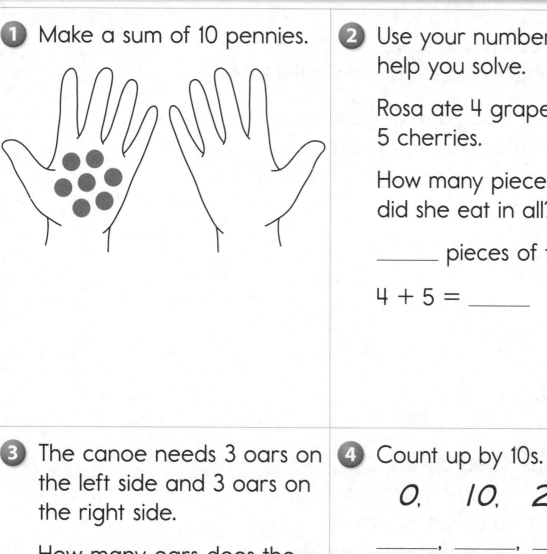

2. Use your number line to help you solve.

   Rosa ate 4 grapes and 5 cherries.

   How many pieces of fruit did she eat in all?

   _____ pieces of fruit

   $4 + 5 =$ _____

3. The canoe needs 3 oars on the left side and 3 oars on the right side.

   How many oars does the canoe need?

   _____ oars

   $3 + 3 =$ _____

4. Count up by 10s.

   0,   10,   20,

   _____, _____, _____,

   _____, _____, _____,

   _____, _____

Introducing Parts-and-Total Diagrams

Read each number story.
Fill in the diagram and solve.
Write a number model. Show your thinking.

1 Lupe read 5 books in September.
She read 4 books in October.
How many books did Lupe read in all?

_____ books

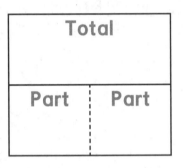

_____ + _____ = _____

2 Emma has 4 red balloons.
She also has some blue balloons.
Emma has 10 balloons total.
How many balloons are blue?

_____ blue balloons

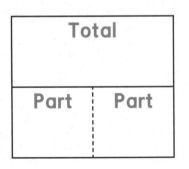

_____ + _____ = _____

Math Boxes

Math Boxes

1 I have 2 tulips and 3 roses. How many flowers do I have in all?

_____ flowers

Number model:

_____ + _____ = _____

2 Find the total.

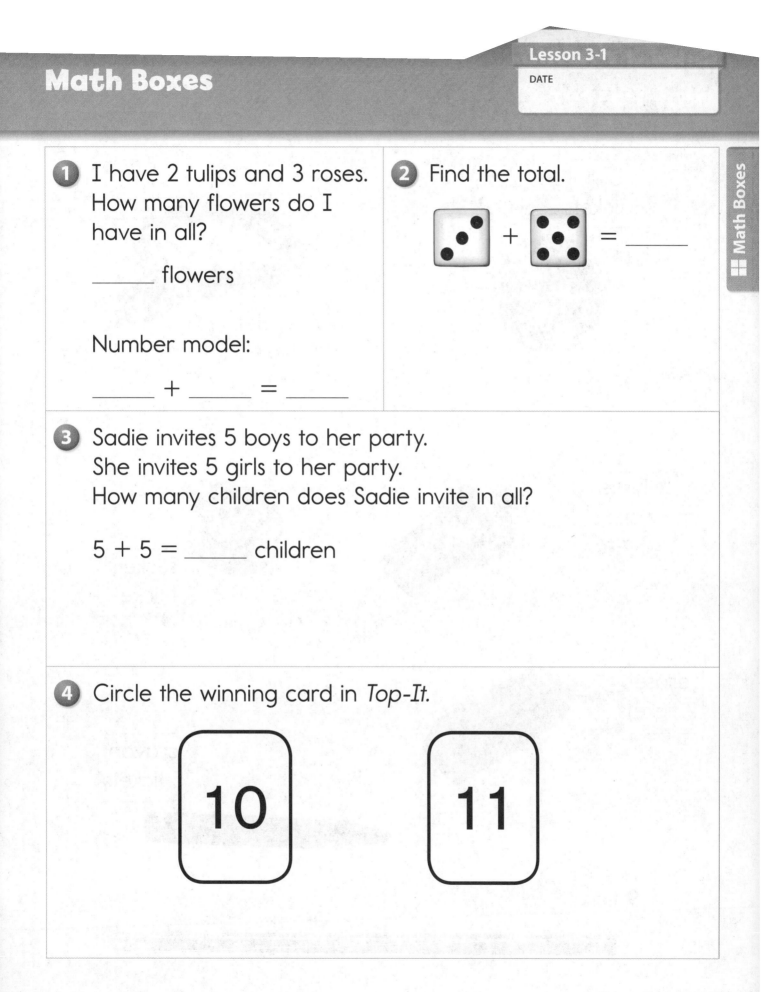

_____

3 Sadie invites 5 boys to her party.
She invites 5 girls to her party.
How many children does Sadie invite in all?

5 + 5 = _____ children

4 Circle the winning card in *Top-It*.

10   11

School Carnival Prizes
Mini-Poster

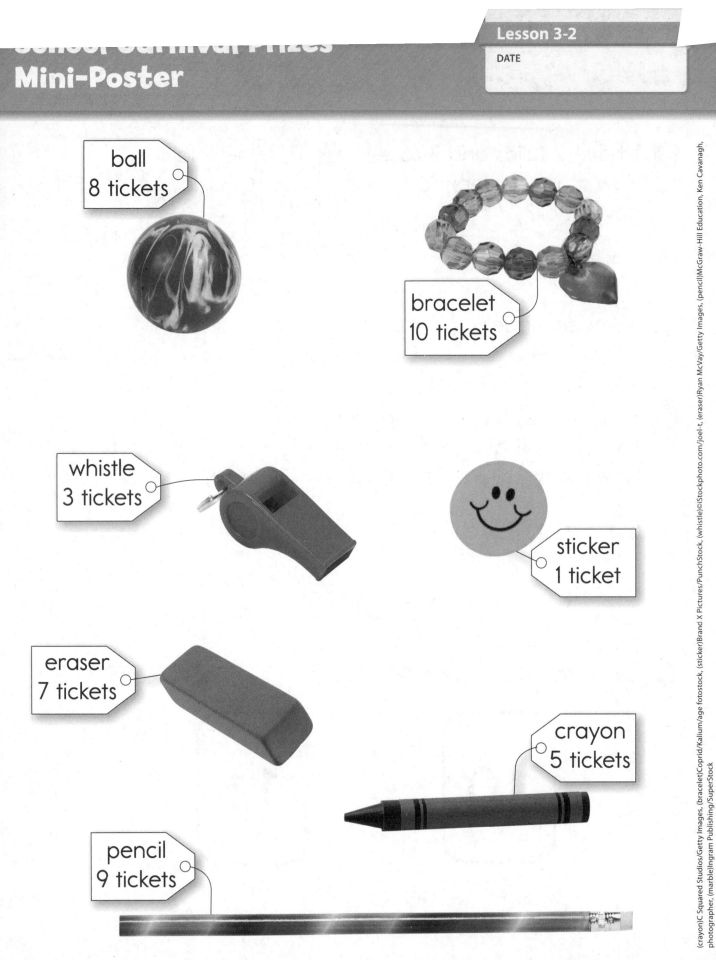

ball
8 tickets

bracelet
10 tickets

whistle
3 tickets

sticker
1 ticket

eraser
7 tickets

crayon
5 tickets

pencil
9 tickets

(crayon)C Squared Studios/Getty Images, (bracelet)Coprid/Kalium/age fotostock, (sticker)Brand X Pictures/PunchStock, (whistle)©iStockphoto.com/joel-t, (eraser)Ryan McVay/Getty Images, (pencil)McGraw-Hill Education, Ken Cavanagh, photographer, (marble)Ingram Publishing/SuperStock

Writing Carnival Number Stories

Write a number story to go with the number sentence.
Then solve.

1 5 + 8 = _____

_____

_____

_____

2 12 − 9 = _____

_____

_____

_____

Try This

3 7 + _____ = 8

_____

_____

_____

Modeling Number Stories

Write a number model to match the story.
Then solve. Use words or pictures to tell
how you solved the problem.

1 David saw 3 ducks in the pond.
Then he saw 2 more ducks on the path.
How many ducks did David see all together?

Number model:

David saw _____ ducks.

2 Sue had 10 rocks.
She gave some rocks to her friend.
Now she has 7 rocks.
How many rocks did Sue give to her friend?

Number model:

Sue gave _____ rocks to her friend.

1 Write a number model for this story.

The gym had 2 baseball bats.
The gym got some new bats.
Now the gym has 5 bats.
How many new bats did the gym get?

Change

Start		End
2	?	5

Unit

bats

_____ + _____ = _____

2 Use your number grid.
Start at 40. Count back 10.
You end at _____.

3 Paz has 8 stickers.
She earns 1 more sticker.
How many stickers does
Paz have now?

Choose the best answer.

◯ 1 sticker
◯ 7 stickers
◯ 8 stickers
◯ 9 stickers

4 **Writing/Reasoning** Solve.

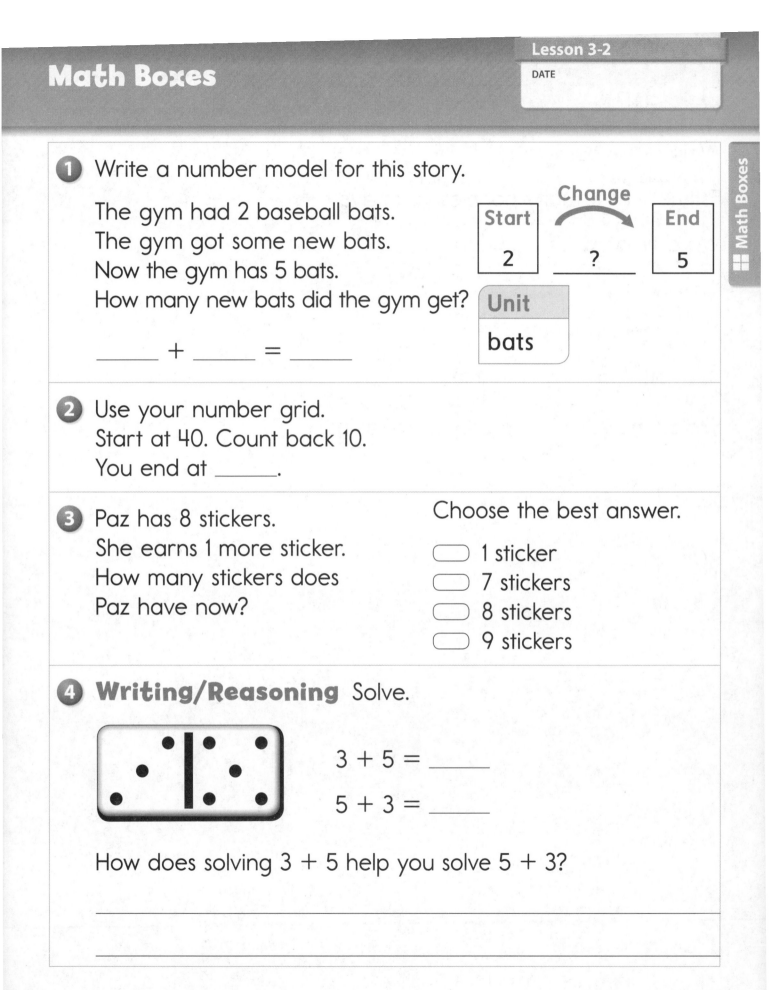

$3 + 5 =$ _____

$5 + 3 =$ _____

How does solving $3 + 5$ help you solve $5 + 3$?

_____

_____

twenty-three   23

Work with your group.

Estimate how many pennies you have.

We estimated _____ pennies.

Now count your pennies.

We counted _____ pennies.

Explain how you counted your pennies.

Matching Pairs

CARD A: ____ + ____ = ____

CARD F: ____ + ____ = ____

CARD B: ____ + ____ = ____

CARD G: ____ + ____ = ____

CARD C: ____ + ____ = ____

CARD H: ____ + ____ = ____

CARD D: ____ + ____ = ____

CARD I: ____ + ____ = ____

CARD E: ____ + ____ = ____

CARD J: ____ + ____ = ____

Math Boxes

Math Boxes

1 Fill in the number model.

Keira has 2 fish and 1 bird. How many pets does Keira have?

_____ + _____ = _____ pets

2 Choose the best answer.

Which die was rolled for this round of *Roll and Total*?

3 + _____ = 7

3 Zak knows 4 songs. Tanya knows 6 songs. How many songs do they know in all?

$4 + 6 =$ _____

$6 + 4 =$ _____

Unit
songs

4 **Writing/Reasoning**

Circle the winning card in *Top-It*.

18   17

How do you know which card has the winning number?

Girls and Boys

Draw a picture of some girls and boys.

How many children did you draw in all? _____ children

Fill in the number model to show what you drew.

_____ girls + _____ boys = _____ children in all

Math Boxes

Math Boxes

1 Count by 5s. Circle your counts on the number grid.

1	2	3	4	5	6	7	8	9	10
11	12	13	14	15	16	17	18	19	20
21	22	23	24	25	26	27	28	29	30
31	32	33	34	35	36	37	38	39	40
41	42	43	44	45	46	47	48	49	50

2 How many ☐s? _____

3 Start at 14. Count up 5.

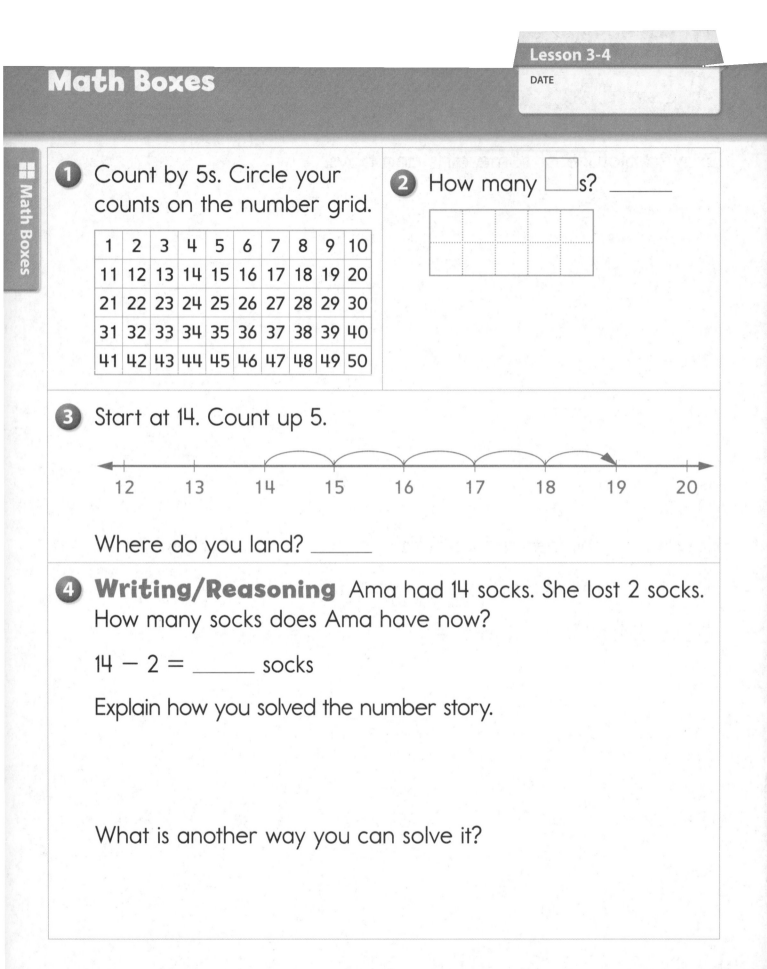

Where do you land? _____

4 **Writing/Reasoning** Ama had 14 socks. She lost 2 socks. How many socks does Ama have now?

14 − 2 = _____ socks

Explain how you solved the number story.

What is another way you can solve it?

Reviewing Skip Counting on Number Lines

1 Show counting by 1s.

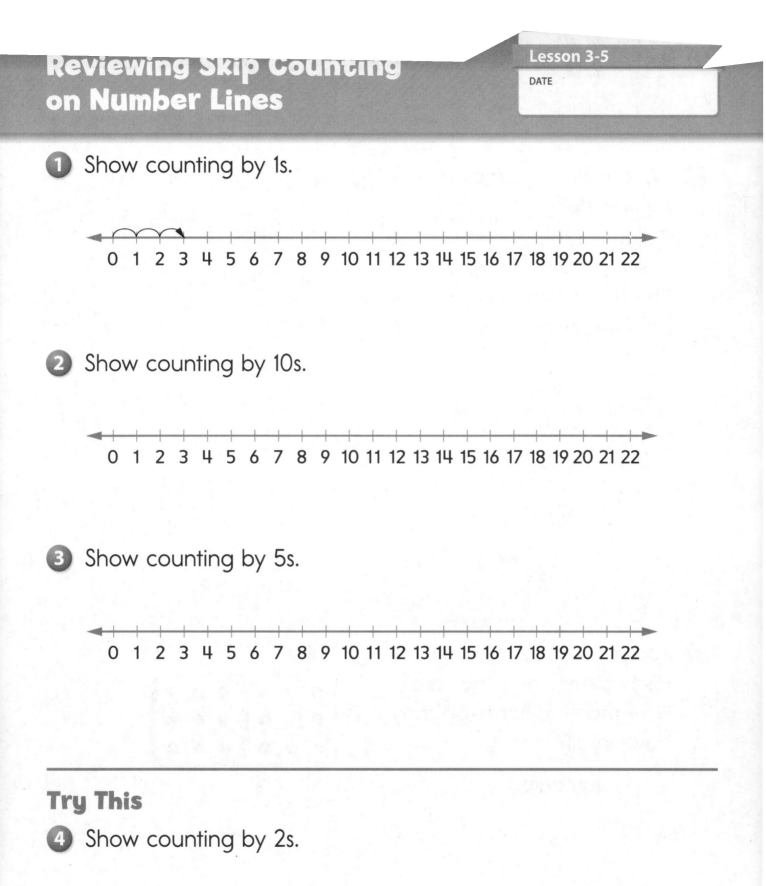

2 Show counting by 10s.

3 Show counting by 5s.

Try This

4 Show counting by 2s.

Math Boxes

Math Boxes

1 Write a number model for the story.

Bella had 10 pennies. She lost some pennies. She has 6 pennies left. How many pennies did Bella lose?

Change

Start		End
10	?	6

_____ − _____ = _____

Unit

2 Use your number grid.

Start at 60.

Count back 30.

You end at _____.

3 Jorge has 3 toy cars. His brother has 7 toy cars. How many toy cars do they have in all?

_____ toy cars

3 + 7 = _____

4 Solve.

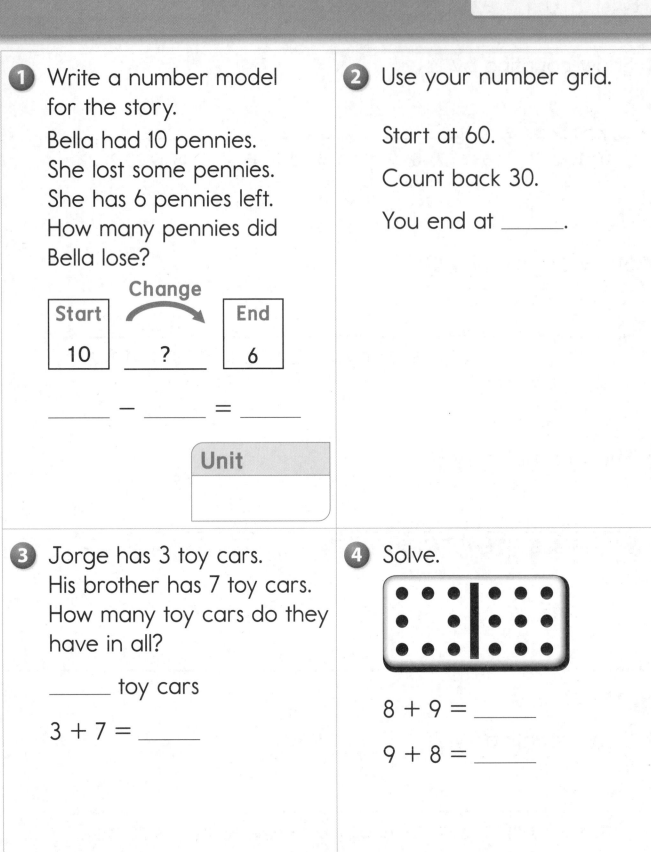

8 + 9 = _____

9 + 8 = _____

Math Boxes

1 Count up by 5s.

__15__, __20__, __25__,

_____, _____, _____,

_____, _____, _____

2 Draw a picture of 12 things.

3 Start at 20. Count back 5.

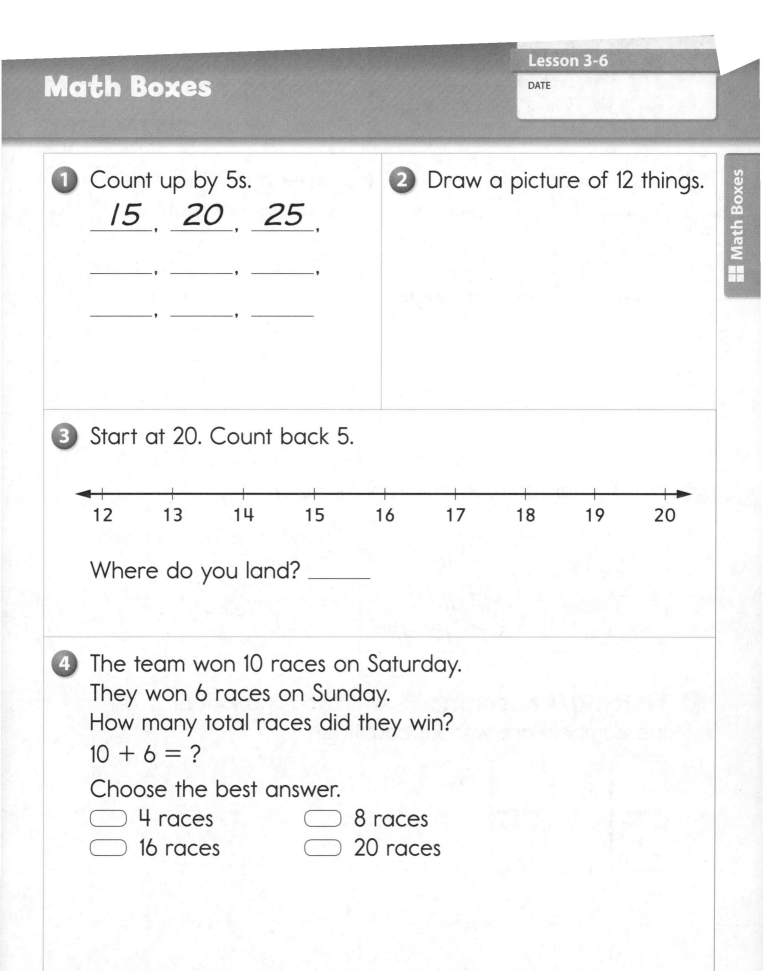

12   13   14   15   16   17   18   19   20

Where do you land? _____

4 The team won 10 races on Saturday.
They won 6 races on Sunday.
How many total races did they win?
$10 + 6 = ?$

Choose the best answer.

⬭ 4 races   ⬭ 8 races
⬭ 16 races   ⬭ 20 races

Math Boxes

1 This is a rectangle.

Draw a different rectangle.

2 Write a number that is greater than 17.

_____

3 Which color do most children like best?

Favorite Colors	
Red	~~HHT~~ ~~HHT~~
Yellow	~~HHT~~ ////
Blue	~~HHT~~ ~~HHT~~ ~~HHT~~

Choose the best answer.

◯ Red
◯ Yellow
◯ Blue

4 **Writing/Reasoning** How do you know what the numbers are without counting?

1 Cady has 3 stripes on each of her two socks. Write a number model to show how many stripes in all.

_____ + _____ = _____

Unit
stripes

2 Use your number grid.
Start on 36.
Count up 10 hops.
Where do you land?

_____

3 Who is taller, you or your teacher?

_____

4 **Writing/Reasoning** Use your number line.
Start on 3.
Count up 5 hops.
Count up 2 more hops.

Where do you land? _____

How is this like starting on 8 and counting up 2 hops?

Solving Frames-and-Arrows Problems

Complete the Frames-and-Arrows diagrams.

1

Rule						
Add 1	6	7				

2

Rule						
Subtract 10	50	40				

3

Rule						
Count back by 1s	20	19			16	

4

Rule						
+ 10	10	20	30			

5

Rule					
Count up by 5s	15	20	25		

Finding Unknown Hops

Write your answers in the ☐.
Use the number line if you like.

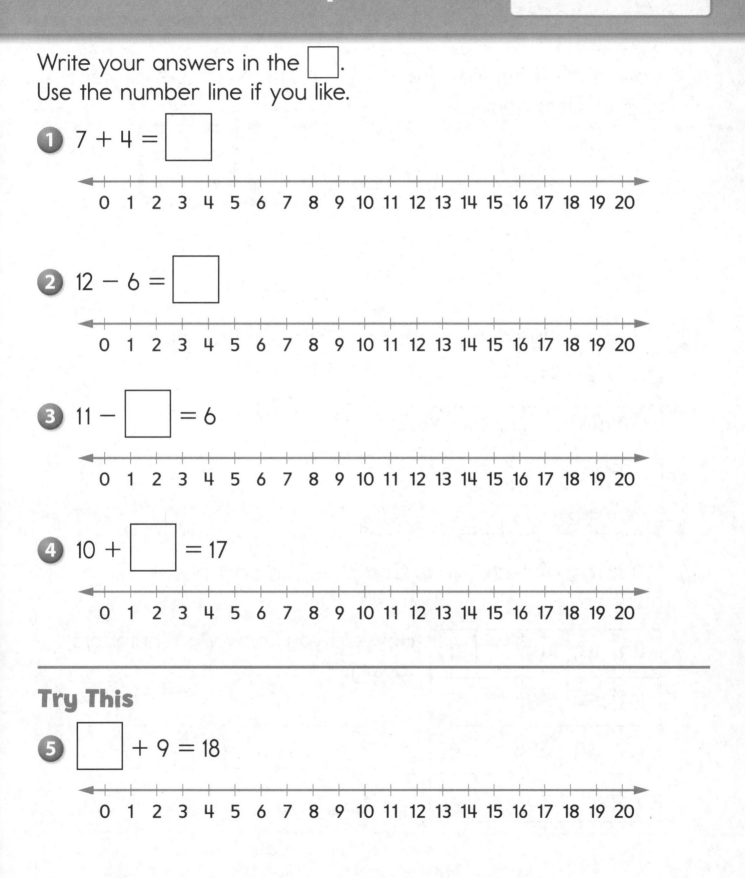

1 $7 + 4 =$ ☐

0 1 2 3 4 5 6 7 8 9 10 11 12 13 14 15 16 17 18 19 20

2 $12 - 6 =$ ☐

0 1 2 3 4 5 6 7 8 9 10 11 12 13 14 15 16 17 18 19 20

3 $11 - $ ☐ $= 6$

0 1 2 3 4 5 6 7 8 9 10 11 12 13 14 15 16 17 18 19 20

4 $10 + $ ☐ $= 17$

0 1 2 3 4 5 6 7 8 9 10 11 12 13 14 15 16 17 18 19 20

Try This

5 ☐ $+ 9 = 18$

0 1 2 3 4 5 6 7 8 9 10 11 12 13 14 15 16 17 18 19 20

thirty-five **35**

Math Boxes

1 How many shoelaces are in your classroom?

_____ shoelaces

2 Fill in the number sentences.

_____ + 6 = 10

10 − 6 = _____

3 Camille counted 15 bushes and 4 trees in her yard.

Complete the tally chart.

Plants in Camille's Yard	
Bushes	
Trees	

4 **Writing/Reasoning** Complete this part of the number grid.

43	44	45		47
53	54	55		57
63	64	65		67
73	74	75		77

How did you know what numbers to write?

_____

_____

_____

_____

Finding Rules in Frames-and-Arrows Problems

1 Fill in the frames.

Rule						
Add 5	15			30		

2 Fill in the rule.

Rule						
	20	18	16	14	12	10

3 Fill in the rule and the frames.

Rule						
	7	10	13			

4 Fill in the rule and the frames.

Rule						
	90	80	70			

5 Make up your own.

Rule						

Adding on the Number Grid

									0
1	2	3	4	5	6	7	8	9	10
11	12	13	14	15	16	17	18	19	20
21	22	23	24	25	26	27	28	29	30
31	32	33	34	35	36	37	38	39	40
41	42	43	44	45	46	47	48	49	50
51	52	53	54	55	56	57	58	59	60
61	62	63	64	65	66	67	68	69	70

1. Start at 25. Count up 3. Where do you end up?

   25 + 3 = _____

2. Start at 19. Count up 6. Where do you end up?

   19 + 6 = _____

3. Start at 38. Count up 2. Where do you end up?

   38 + 2 = _____

4. Start at 57. Count up 10. Where do you end up?

   57 + 10 = _____

5. Start at 62. Count up 0. Where do you end up?

   62 + 0 = _____

Math Boxes

1 Draw 2 different squares.

2 Write a number less than 26.

_____

3

December Weather	
Sunny	////-////-//
Rainy	////-////
Snowy	////-////

How many days were sunny?

_____ sunny days

Were there more rainy days or snowy days?

More _____ days

4 Write the number.

____   ____

thirty-nine   39

Math Boxes

Ordering by Length

1 Circle the shortest pencil.
Make an X on the longest pencil.

2 Circle the shortest vehicle.
Make an X on the longest vehicle.

Try This

3 Put the children in order from shortest to tallest.

Child _____ is shortest.

Child _____ is in the middle.

Child _____ is tallest.

1 How many buildings can you see outside your classroom window?

_____ buildings

2

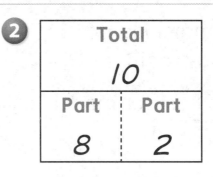

Total
10

Part	Part
8	2

$8 + 2 = 10$

Which subtraction sentence matches this diagram?

Choose the best answer.

◯ $10 - 8 = 2$
◯ $8 - 2 = 10$
◯ $9 - 2 = 8$
◯ $10 - 2 = 6$

3 Addy counted cars. She counted 10 white cars, 8 silver cars, and 9 black cars. Complete the tally chart.

Cars	
White	
Silver	
Black	

4 Complete this part of the number grid.

4	5	6	7	
14	15	16	17	
24	25	26	27	
34	35	36	37	
44	45	46	47	

Math Boxes
Preview for Unit 4

Math Boxes

1 Find a matching pair in your classroom.

Write a number model for it.

☐ + ☐ = ☐

2 Use your number grid.

Start on 50.
Count back 10 hops.
Where do you land?

_____

3 Write the name of a classmate.

_____

Who is taller, you or your classmate?

_____

4 Use your number line.

Start on 6.
Count up 3 hops.
Count up 5 more hops.
Where do you land?

_____

Introducing Length Comparison

Write or draw things that are **a bit shorter** than your paper.

Things that are **a bit shorter** than my paper:

Write or draw things that are **a bit longer** than your paper.

Things that are **a bit longer** than my paper:

Are the things in the top box **longer** or **shorter** than
the things in the bottom box?

_____

_____

Tell your partner how you know.

Subtracting on a Number Grid

-9	-8	-7	-6	-5	-4	-3	-2	-1	0
1	2	3	4	5	6	7	8	9	10
11	12	13	14	15	16	17	18	19	20
21	22	23	24	25	26	27	28	29	30
31	32	33	34	35	36	37	38	39	40
41	42	43	44	45	46	47	48	49	50
51	52	53	54	55	56	57	58	59	60
61	62	63	64	65	66	67	68	69	70

1 Start at 38. Count back 4. Where do you land? _____

2 Start at 53. Count back 6. Where do you land? _____

3 Start at 60. Count back 10. Where do you land? _____

4 Start at 40. Count back 30. Where do you land? _____

Try This

Solve.

5 Start at 52. Count back 20. Where do you land? _____

6 Start at 47. Count back 17. Where do you land? _____

Math Boxes

1 Use your number grid.
Start at 8.
Count up _____ hops.
You end at 12.

Choose the best answer.

○ 2 ○ 4
○ 8 ○ 10

2 Draw the missing dots.
Fill in the number sentence.

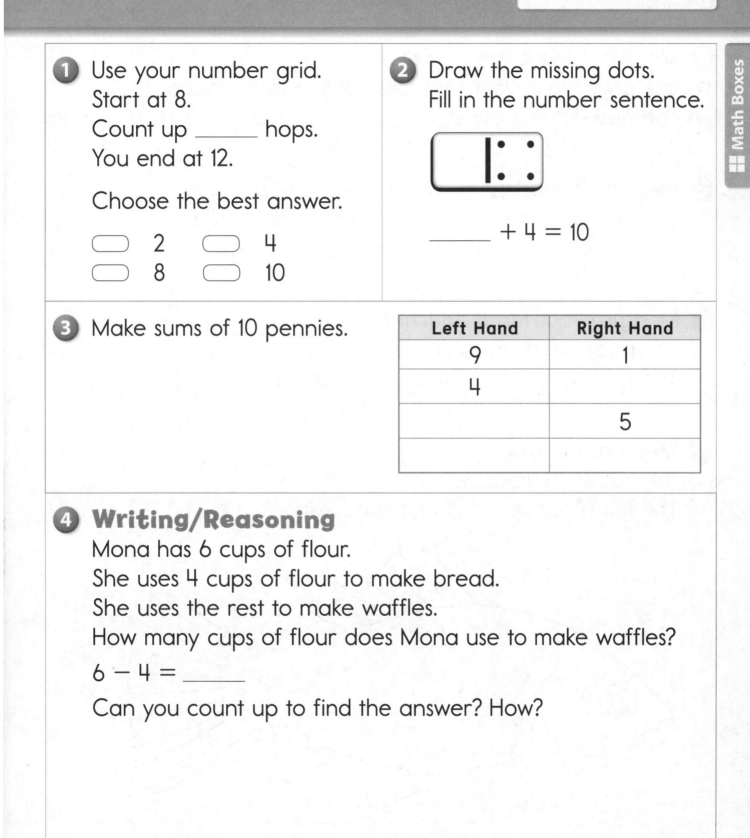

_____ + 4 = 10

3 Make sums of 10 pennies.

Left Hand	Right Hand
9	1
4	
	5

4 **Writing/Reasoning**
Mona has 6 cups of flour.
She uses 4 cups of flour to make bread.
She uses the rest to make waffles.
How many cups of flour does Mona use to make waffles?

6 − 4 = _____

Can you count up to find the answer? How?

Math Boxes

Measuring Length

1 Measure your **desk** or **table**.
Use anything in your measurement kit to measure.
The desk/table is about _____ long.

2 Measure a **marker**.
Use anything in your measurement kit to measure.
The marker is about _____ long.

3 Measure your *Math Journal.*
Use **paper clips** to measure.
The *Math Journal* is about _____ paper clips long.

4 Measure the **fish**.
Use **cubes** to measure.
The fish is about _____ unit cubes long.

Math Boxes

1 Use your number line.
Start on 3.
Count up 6.
Where do you land?

$3 + 6 =$ ☐

2 Use your number line.
Solve.

$4 + \underline{\hspace{1cm}} = 6$

$6 - 4 = \underline{\hspace{1cm}}$

Can you count up to solve both number sentences?

3 Write the sums.

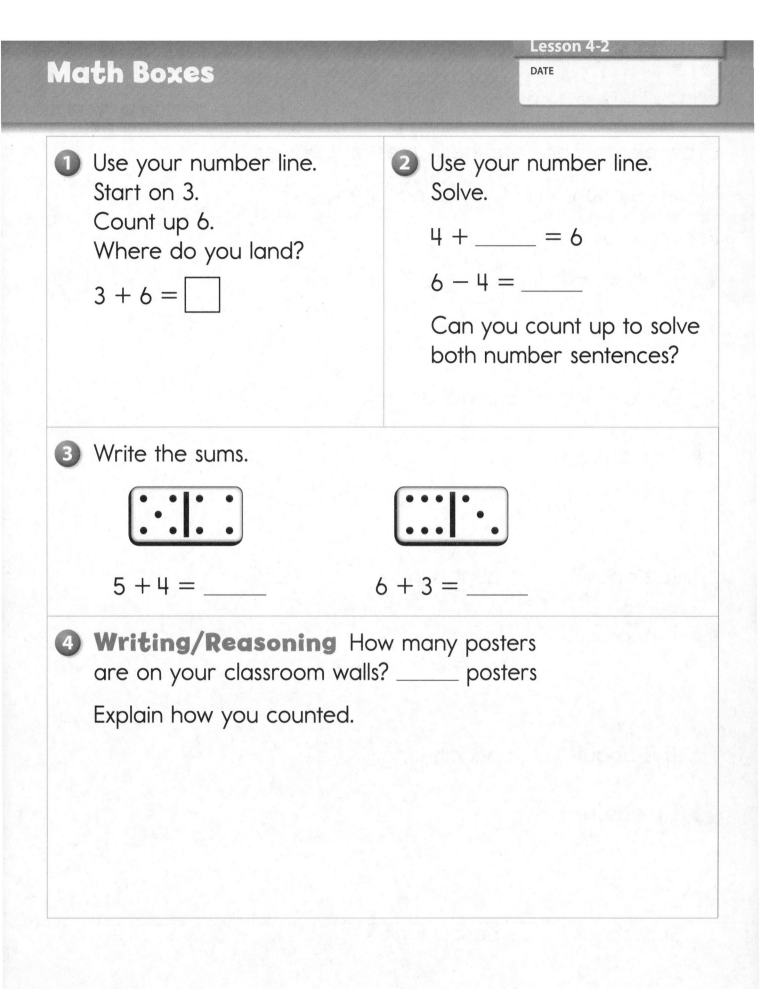

$5 + 4 = \underline{\hspace{1cm}}$ $6 + 3 = \underline{\hspace{1cm}}$

4 **Writing/Reasoning** How many posters are on your classroom walls? $\underline{\hspace{1cm}}$ posters

Explain how you counted.

Measuring with Pencils

Draw pictures or write the names of 4 objects.

Measure each object with a new pencil.

Record your answer.

1 I measured

It is about _____ pencils long.

2 I measured

It is about _____ pencils long.

3 I measured

It is about _____ pencils long.

4 I measured

It is about _____ pencils long.

Practicing Comparing Numbers

1 Circle the smaller number in each pair.

14   23

27   32

2 Circle the larger number in each pair.

79   66

34   42

3 Circle the smaller number in each pair.

14   13

29   31

18   22

4 Circle the larger number in each pair.

64   54

73   77

45   54

Try This

5 Write the numbers in order from smallest to largest.

28   54   7   23   12   45

____   ____   ____   ____   ____   ____

Math Boxes

Math Boxes

1 Use your number grid.
Start at 3.
How many more hops to get 11?

$3 + \boxed{} = 11$

2 Draw the missing dots. Fill in the number sentence.

_____ $+ 6 = 14$ _____ $+ 9 = 13$

3 Make sums of 10 pennies.

Left Hand	Right Hand
3	7
6	
	8

4 Freddie has 12 pencils.
He gives 5 pencils to his friends.
How many pencils does he have left?

Change

Start		End
12	5	?

$12 - 5 =$ _____

Can you count up to solve? _____

1 Which ribbon is longer?

Circle it.

2 Which set of tally marks shows the larger number?

Circle it.

卌 卌 I

卌 III

Math Boxes

Math Boxes

1. Fill in the rule and write the missing numbers.

   Rule

   8   10   12   ☐   ☐

2. Write the missing numbers.

   Rule
   −6

   26   20   ☐   ☐   2

3. Use your number grid.

   Start at 45.
   Count up 10 hops.
   Where do you land? _____

4. **Writing/Reasoning** How many dots?

   _____ dots

   Explain how you can tell how many without counting.

Measuring with Paper Clips

1 Circle the better measurement.

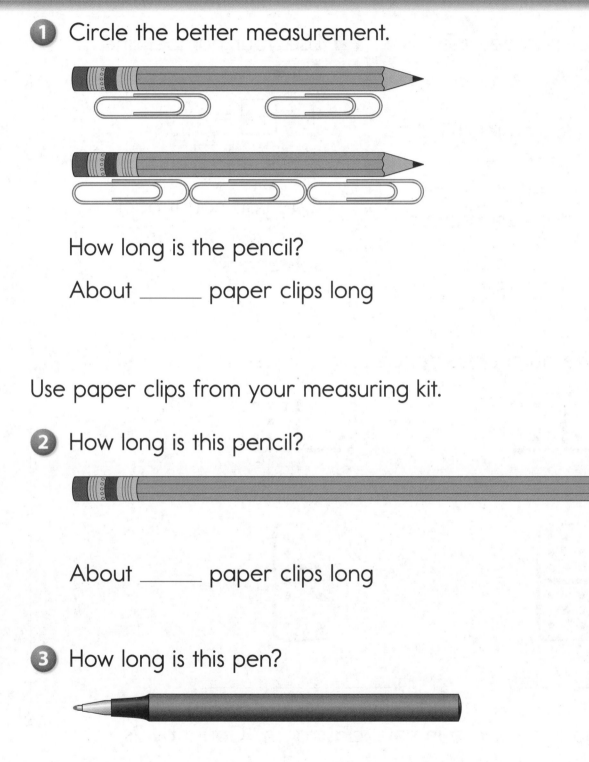

How long is the pencil?

About _____ paper clips long

Use paper clips from your measuring kit.

2 How long is this pencil?

About _____ paper clips long

3 How long is this pen?

About _____ paper clips long

Math Boxes

1 Use your number line.
Start on 7.
Count up 0.
Where do you land?

$7 + 0 = $ ☐

Choose the best answer.

◯ 1   ◯ 6
◯ 7   ◯ 8

2 Use your number line.
Solve.

$4 + $ ☐ $= 8$

$8 - 4 = $ ☐

Can you count back to solve both number sentences?

_____

3 Write the number sentences.

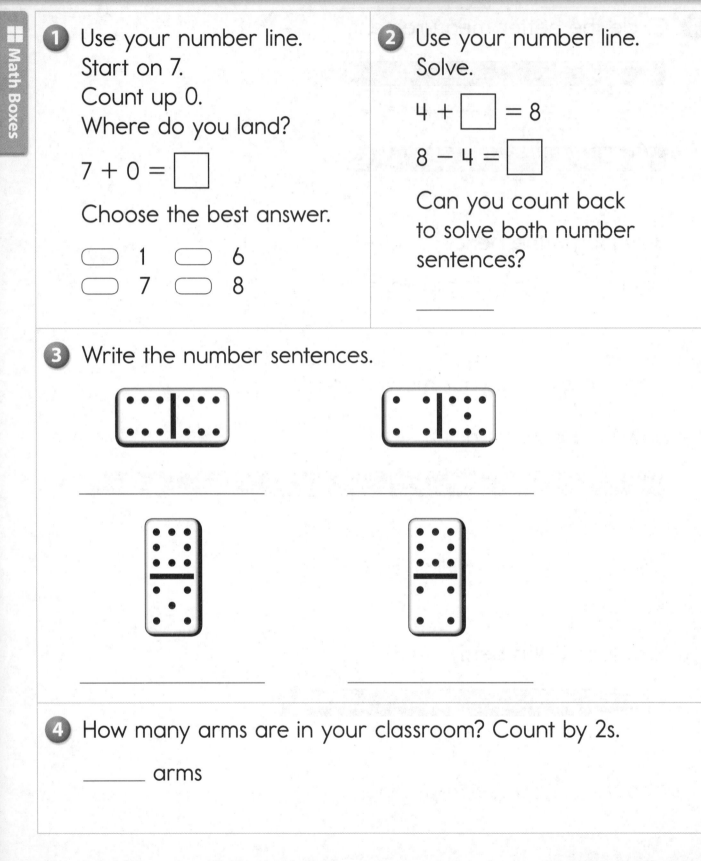

_____   _____

_____   _____

4 How many arms are in your classroom? Count by 2s.

_____ arms

Writing and Solving Number Models

Write a number model for each number story. Solve.
Use diagrams if you like.

1 10 marbles were on the table.
6 marbles rolled onto the floor.
How many marbles were left on the table?

Number model: _____

_____ marbles

2 14 birds were on a fence.
Some of the birds flew away.
Now 11 birds are on the fence.
How many birds flew away?

Number model: _____

_____ birds

3 Danielle brought 16 bananas to school.
Each child in her class took a banana.
There were 2 bananas left.
How many children took a banana?

Number model: _____

_____ children

Math Boxes

Math Boxes

1 What is the rule?

Rule ?

70 60 50 40 30

Choose the best answer.

- ⬭ Count up by 2s
- ⬭ + 10
- ⬭ Subtract 5
- ⬭ − 10

2 Write the missing numbers.

Rule +3

0 ( ) ( ) 9 ( )

3 Use your number grid.

Start at 10.
Count up 13 hops.

Where do you land?

_____

4 How many dots in all?

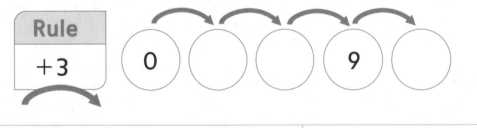

_____ dots   _____ dots

Math Boxes

Doubles

Think of numbers you can add together that you

- know would make a double,

- know would NOT make a double, and

- are not sure about.

Write them in the columns.

Double	NOT Double	Not Sure
Example: 1 + 1	Example: 1 + 3	Example: 13 + 31

Math Boxes

Math Boxes

1 Use a number grid.
Count by 10s.

_8_ , _18_ , _____ , _____ ,

_____ , _____ , _____ , _____ ,

_____ , _____ , _____

2 Write a number that is more than 39.

_____

3 Solve.

Some sloths have 2 toes on each front foot.

How many toes on their front feet in all? _____ toes

Other sloths have 3 toes on each front foot.

How many toes on their front feet in all? _____ toes

4 Draw and solve.

Olga had 6 pennies.

Tyson gave her 2 more pennies.

How many pennies does Olga have now?

_____ pennies

1 How many pennies could you exchange for 1 dime?

_____ pennies

2 Start at 40.

Count back 20.

You end at _____.

3 Subtract.

$80 - 20 =$ ☐

$70 - 30 =$ ☐

$50 - 40 =$ ☐

Unit

4 **Writing/Reasoning** Which path is longer? Circle your answer.

The path from your seat to the door

OR

The path from your seat to the bathroom

How do you know?

Measuring with Erasers

1 Circle the better measurement.

How long is the truck?

About _____ erasers

2 Neesha used erasers to measure some things.

How long are the scissors?

About _____ erasers

How long is the necklace?

About _____ erasers

Math Boxes

1 Solve.

$1 + 1 =$ ☐

$2 + 2 =$ ☐

☐ $- 3 = 3$

$8 -$ ☐ $= 4$

Unit
fish

2 Draw and solve.

The garden has 4 ladybugs and 10 ants.
How many insects are there in all?

_____ insects

3 How many pennies wide is this page?

About _____ pennies

4 **Writing/Reasoning** Solve.
Nadia has 5 stars.
Cathy has 8 stars.
How many more stars does Cathy have than Nadia?

$5 +$ _____ $= 8$ stars

How could you count up to find the difference?

Write a number story to go with the number sentence.
Then solve.

1 $2 + 3 + 5 =$ _____

2 $4 + 6 + 8 =$ _____

3 $3 + 9 + 1 =$ _____

Creating School Supply Number Stories 2

Write a number story to go with the number sentence.
Then solve.

① $7 + 3 + 4 =$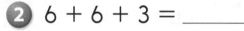

② $6 + 6 + 3 =$

③ $9 + 2 + 4 =$

Math Boxes

1 Use a number grid.
Count by 10s.

_____/_____, _____//_____, _____, _____,

_____, _____, _____, _____,

_____, _____, _____

2 Which number is less than 23?

Choose the best answer.

⊂⊃ 24   ⊂⊃ 31
⊂⊃ 50   ⊂⊃ 14

3 Solve.

$10 + \underline{\qquad} = 20$

$4 + \boxed{} = 8$

$6 = 3 + \underline{\qquad}$

$14 = \boxed{} + 7$

Unit
bugs

4 **Writing/Reasoning** Draw and solve.

Ava had 9 pennies. She lost 4 pennies.
How many pennies does Ava have now?

_____ pennies

How does drawing a picture help you solve a number story?

Solving More Carnival Number Stories

Use the School Carnival Prizes Mini-Poster on page 20.
Solve. Then write a number model for each number story.

1 David wants a pencil and a sticker.

How many tickets does he need? _____ tickets

Number model: _____

2 Christopher has 3 tickets.

He wants an eraser.

How many more tickets does he need

to get the eraser? _____ tickets

Number model: _____

3 Jada wants a ball.

She paid with 9 tickets.

How many tickets will she get back? _____ ticket

Number model: _____

Try This

4 Write a number story to go with the number sentence. Solve.

$10 - 7 =$ _____

_____

_____

1 Solve.

$3 - 0 = \boxed{}$

$3 + \boxed{} = 4$

$\begin{array}{r} \boxed{} \\ + 0 \\ \hline 6 \end{array}$

$\begin{array}{r} 6 \\ - 1 \\ \hline \boxed{} \end{array}$

Unit

2 A chicken laid 12 eggs.
2 eggs hatched.
How many eggs did not hatch?

Choose the best answer.

- ⬭ 2 eggs
- ⬭ 10 eggs
- ⬭ 11 eggs
- ⬭ 14 eggs

3 How many pennies long is this page?

About _____ pennies

4 Ashur's class must run 4 laps around the gym.
Ashur has run 2 laps so far.
How many more laps does he have to run?

$2 + \boxed{} = 4$

Unit

laps

1. How many dimes can you exchange for 20 pennies?

   _____ dimes

2. Start at 33.
   Count back 10.
   Where do you end?
   Choose the best answer.

   ◯ 38
   ◯ 46
   ◯ 23
   ◯ 83

3. Solve.

   $60 - \boxed{} = 50$

   $90 - 30 = \boxed{}$

   $\boxed{} - 10 = 20$

   Unit

4. Which path is longer?
   Circle your answer.

   The path from your classroom door to the playground

   The path from your classroom door to the office

Tens-and-Ones Riddles

Solve the riddles. Use base-10 blocks to help you.

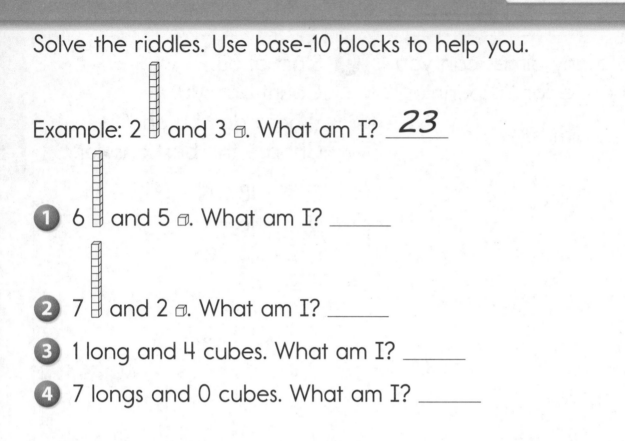

Example: 2 ▌ and 3 ▫. What am I? __23__

1 6 ▌ and 5 ▫. What am I? _____

2 7 ▌ and 2 ▫. What am I? _____

3 1 long and 4 cubes. What am I? _____

4 7 longs and 0 cubes. What am I? _____

Try This

Trade to find the answers.

5 1 long and 11 cubes. What am I? _____

6 2 longs and 14 cubes. What am I? _____

7 Use your blocks to make 16 in two ways.

Draw your ways using ▌ and ▪.

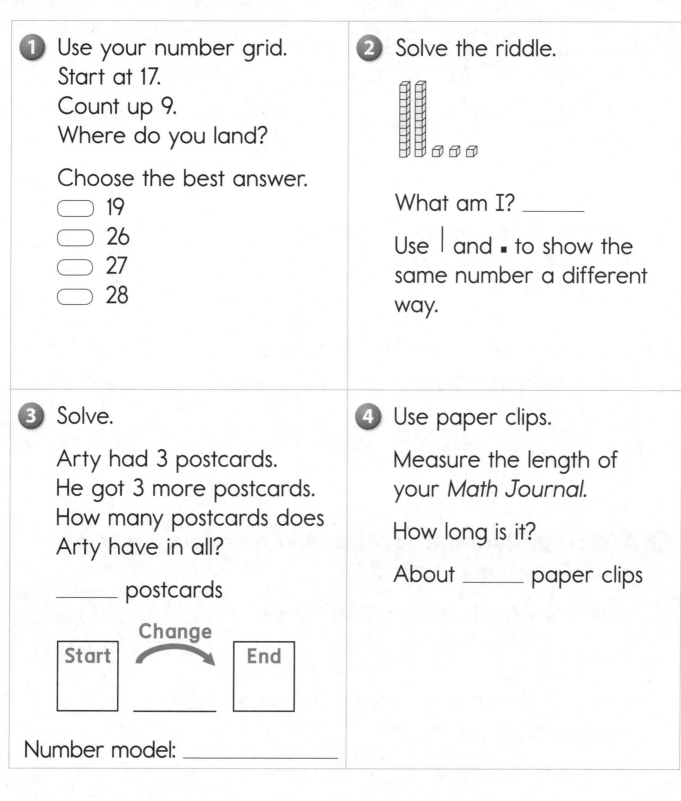

1 Use your number grid.
Start at 17.
Count up 9.
Where do you land?

Choose the best answer.
- ⬭ 19
- ⬭ 26
- ⬭ 27
- ⬭ 28

2 Solve the riddle.

What am I? _____

Use | and ▪ to show the same number a different way.

3 Solve.

Arty had 3 postcards.
He got 3 more postcards.
How many postcards does
Arty have in all?

_____ postcards

Change

| Start | → | End |

_____

Number model: _____

4 Use paper clips.

Measure the length of your *Math Journal.*

How long is it?

About _____ paper clips

Math Boxes

DATE

1

Weather	
Sunny	~~HHH~~ ~~HHH~~ ~~HHH~~ ///
Rainy	~~HHH~~ ~~HHH~~
Snowy	~~HHH~~ ////

How many sunny days?

_____ sunny days

Were there more rainy days or snowy days?

More _____ days

2 Sadie drew 9 stars and 7 moons.
How many shapes did Sadie draw in all?

9 + 7 = _____

Unit

shapes

3 Use your number grid. Which number is 10 more than 78?
Choose the best answer.

⬭ 79
⬭ 80
⬭ 87
⬭ 88

4 **Writing/Reasoning** Abby and Tyler have 7 markers in all. Abby has 4 markers.

How many markers does Tyler have?

7 − 4 = ☐

Unit

markers

What addition number model could you use to solve this subtraction problem?

_____

Use another object to compare the lengths of these objects.

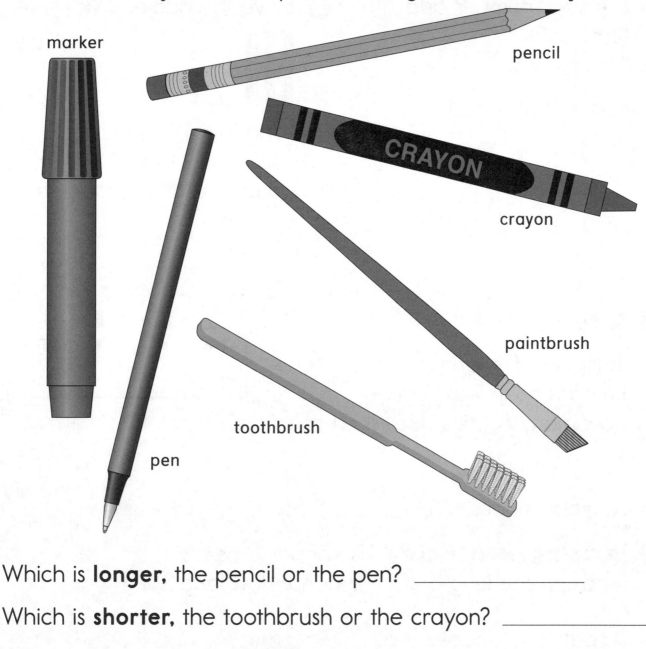

marker

pencil

CRAYON

crayon

paintbrush

toothbrush

pen

Which is **longer,** the pencil or the pen? _____

Which is **shorter,** the toothbrush or the crayon? _____

Which object is the **longest?** _____

Which object is the **shortest?** _____

What did you use to compare? _____

Show a partner how you compared the lengths.

Math Boxes

Math Boxes

1 Use your number grid.
Start at 14.
Count up 6.

You end at _____.

2 Solve the riddle.

What am I? _____
Use ❙ and ∎ to show this
number a different way.

3 Solve.

Betsy had 6 cousins.
1 more cousin was born.
How many cousins does she have now?

_____ cousins

Number model: _____

Change

Start		End
	_____	

4 **Writing/Reasoning** Use paper clips.
Measure the length of your hand. How long is it?

About _____ paper clips

Tell how you measured your hand.

_____

_____

Math Boxes

1 Solve.

Andi used her number line.
She moved up 6 hops and landed on 12.

Where did she start? _____

_____ + 6 = 12

2 Complete the bar graph.

Alice earned 3 stars.
Gabe earned 4 stars.
Chloe earned 1 star.

Stars Earned

Number
of
Stars

4		
3		
2		
1		

Alice   Gabe   Chloe

3 Draw and solve.

Yuko has 3 red balloons, 4 green balloons, and 1 blue balloon. How many balloons does she have in all?

_____ balloons

4 **Writing/Reasoning**
Circle the winning roll in *Roll and Total*.

How do you know who won?

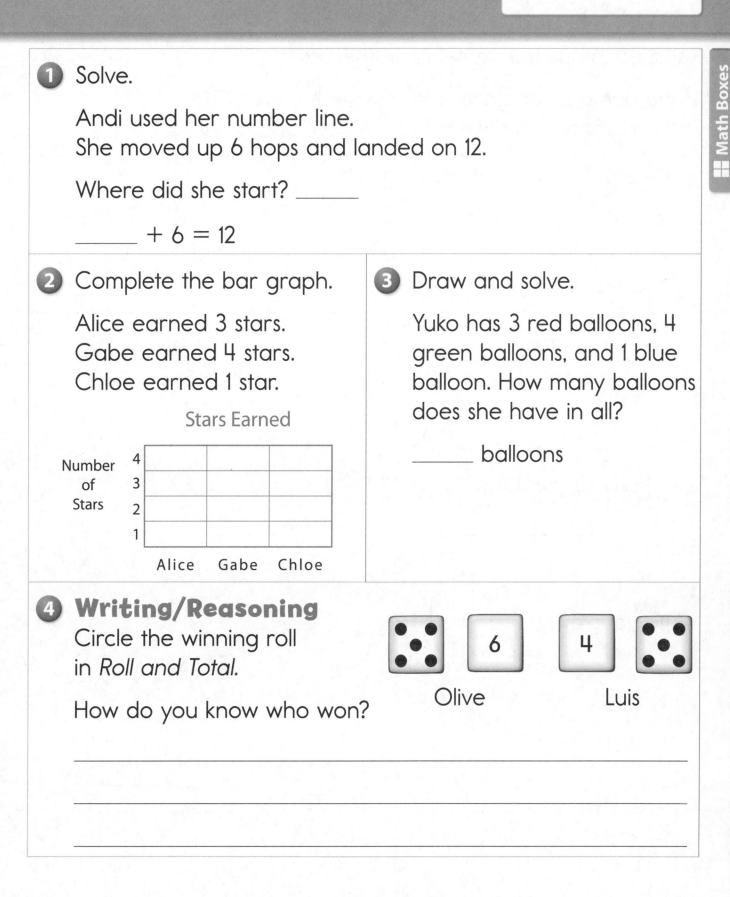

Olive   Luis

_____

_____

_____

True and False
Number Sentences

Read each number sentence in the box.

If the number sentence is true, write it under TRUE.
If the number sentence is false, write it under FALSE.

12 = 12	5 = 10 − 5	10 + 6 = 15
18 = 8 + 10	5 = 13	3 + 1 = 2 + 2
13 − 1 = 14	7 + 8 = 14	6 + 6 = 13
5 + 2 = 2 + 5		

TRUE   **FALSE**

_____   _____

_____   _____

_____   _____

_____   _____

_____   _____

Write your own!

TRUE   **FALSE**

_____   _____

_____   _____

_____   _____

Comparing Numbers Using >, <, and =

Write the numbers shown with base-10 blocks.
Decide which number is larger.
Use <, >, or = to compare the numbers.

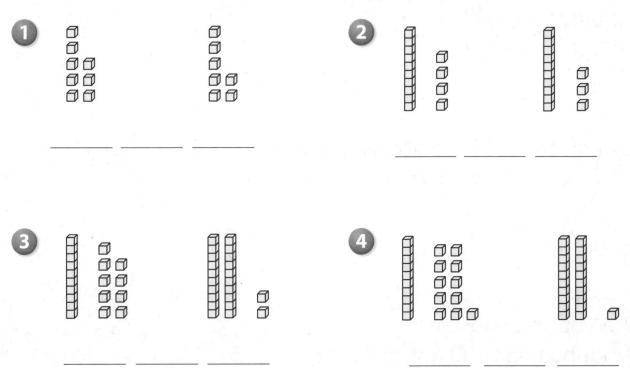

① _____ _____ _____

② _____ _____ _____

③ _____ _____ _____

④ _____ _____ _____

⑤ Compare the numbers. Use <, >, or =.

9 _____ 7   20 _____ 10

18 _____ 27   5 _____ 15

1

Weather	
Sunny	~~HHH~~ ~~HHH~~ ~~HHH~~ ///
Rainy	~~HHH~~ ~~HHH~~
Snowy	~~HHH~~ ////

How many more sunny days were there than rainy days?

_____ days

2 Owen read 8 sports stories and 5 dog stories. How many stories did Owen read in all?

$8 + 5 =$ _____

Unit

3 Use your number grid.
What number is 10 less than 26?

_____

4 Eli and Ava have 10 counters in all.
Eli has 8 counters.
How many counters does Ava have?

_____ counters

$10 - 8 = \boxed{}$

$8 + \boxed{} = 10$

Math Boxes

Math Boxes

1. Use your number grid.
Start at 15.
Count up 7.

You end at _____.

2. How much money?

Ⓓ Ⓓ Ⓟ Ⓟ Ⓟ Ⓟ Ⓟ Ⓟ Ⓟ Ⓟ Ⓟ Ⓟ

_____ cents

Make a trade and show this amount with different coins.
Use Ⓟ and Ⓓ.

3. Solve.

Mai had 5 toy cars. She found 4 more toy cars.

How many toy cars does Mai have in all? _____

Number model: _____

Unit

4. Use your pinky finger.

Measure one side of your desk. How long is it?

About _____ pinky fingers

Make two different paths using 3 paper strips.
Draw your paths below.
Measure each path with one paper clip.
Label each drawing.
Write number models that show the length of each path.

PATH 1:

My path is _____ paper clips long.

Number model: _____

Check by placing paper clips along the whole path.

Did you get the same length? _____

PATH 2:

My path is _____ paper clips long.

Number model: _____

Check by placing paper clips along the whole path.

Did you get the same length? _____

Practicing Addition Facts

Solve.

1 _____ + 5 = 5   10 = _____ + 3   16 = 8 + _____

2 _____ + 4 = 10   10 = 4 + _____   _____ = 6 + 6

3
$$\begin{array}{r} 8 \\ + \ 7 \\ \hline \end{array}$$
$$\begin{array}{r} 6 \\ + \ 3 \\ \hline \end{array}$$
$$\begin{array}{r} 4 \\ + \ 4 \\ \hline \end{array}$$

4 3 + _____ = 10   5 + 5 = _____   8 + _____ = 8

5 _____ + 10 = 10   10 = 1 + _____   2 + _____ = 9

Try This

6 999 + 0 = _____

7 Explain how you solved Problem 6.

_____

_____

_____

Math Boxes

DATE

Math Boxes

1 Layla used her number line. She moved back 9 hops and landed on 9. Where did she start? _____

$$\boxed{} - 9 = 9$$

2 Fill in the bar graph.
Mia earned 3 stars.
Jay earned 2 stars.
Al earned 0 stars.

Stars Earned

3 Lian has 3 red fish, 6 blue fish, and 7 orange fish. How many fish does he have in all?

Choose the best answer.
- ⬭ 9 fish   ⬭ 13 fish
- ⬭ 14 fish   ⬭ 16 fish

4 Write > or < to show the winning domino in *Domino Top-It.*

Ben

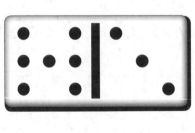

Kat

Finding Paths to the Treasure

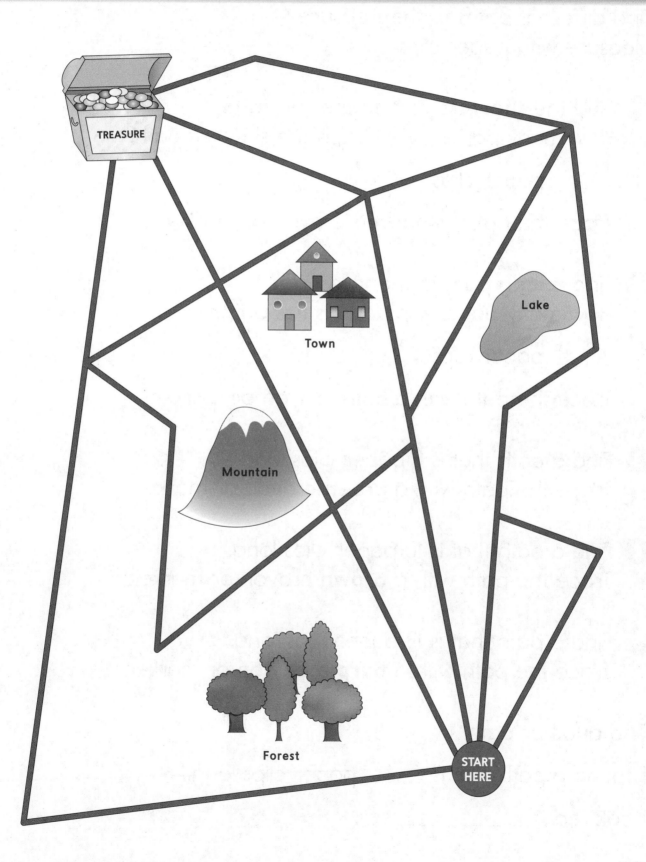

TREASURE

Town

Lake

Mountain

Forest

START HERE

Find different paths to the treasure.
Measure with paper clips.

1 Find the **shortest** path to the treasure.
How many paper clips long is the shortest path?

_____ paper clips

Trace the path with a **red** crayon or marker.

2 Find a **long** path to the treasure.
How many paper clips long is this path?

_____ paper clips

Trace the path with a **blue** crayon or marker.

3 Find a path that is 8 paper clips long.
Trace the path with a **green** crayon or marker.

4 Find a path that is 10 paper clips long.
Trace the path with a **brown** crayon or marker.

5 Find a path that is 13 paper clips long.
Trace the path with a **purple** crayon or marker.

Find another path.

I found a path that is _____ paper clips long.

I colored it _____.

Math Boxes

1 Liam used his number line.

He started at 9. He moved forward 0 hops.

Where did he land?
Choose the best answer.

◯ 0
◯ 8
◯ 9
◯ 10

2 Draw a pencil that is 3 paper clips long.

3 Use your number grid.

Start at 20. Count back 9.

You end at _____.

4 **Writing/Reasoning** How do you know which pet is most popular?

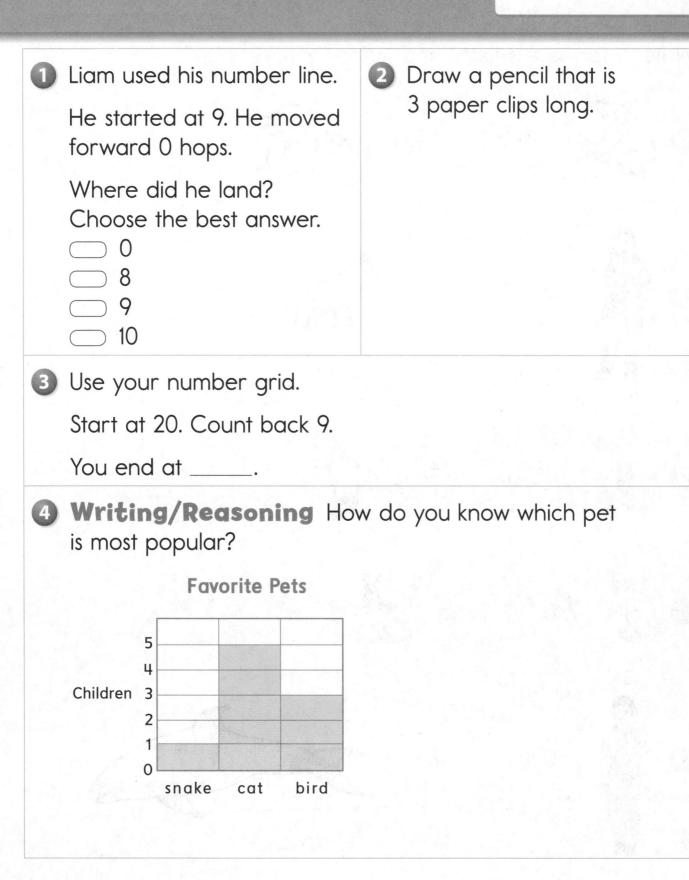

Favorite Pets

Less Than and More Than Number Models

Write < for "is less than" and > for "is more than."

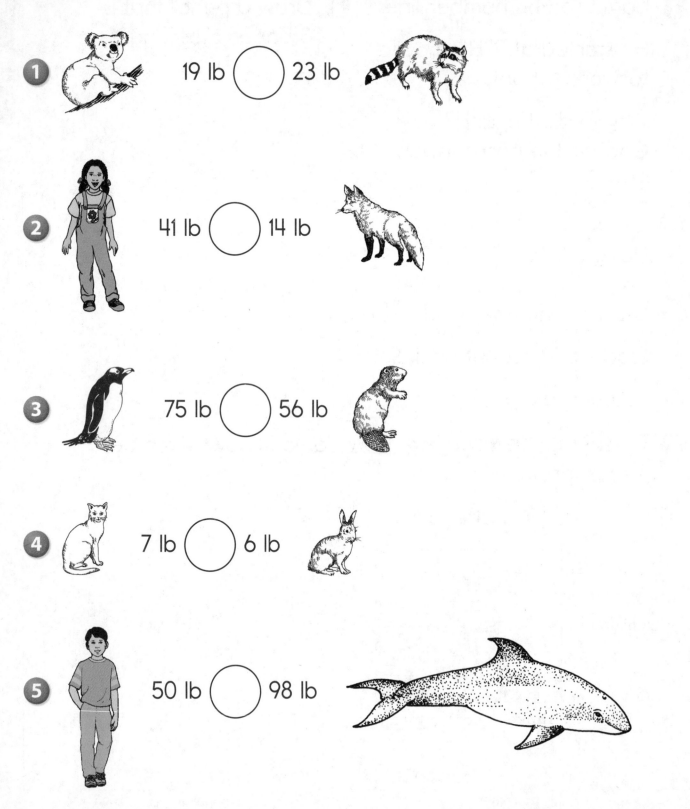

1 19 lb ◯ 23 lb

2 41 lb ◯ 14 lb

3 75 lb ◯ 56 lb

4 7 lb ◯ 6 lb

5 50 lb ◯ 98 lb

"Less Than" and "More Than" Number Models (continued)

Write < for "is less than" and > for "is more than."

6 7 lb + 6 lb ◯ 15 lb

7 19 lb ◯ 14 lb + 3 lb

8 5 lb + 8 lb ◯ 20 lb

How did you find your answer to Problem 8?

_____

_____

_____

_____

Try This

9 75 lb ◯ 56 lb + 20 lb

Preview for Unit 6

Math Boxes

1 Add.

$2 + 2 =$ _____

$3 + 3 =$ _____

$4 + 4 =$ _____

$5 + 5 =$ _____

Unit

2 Record the time.

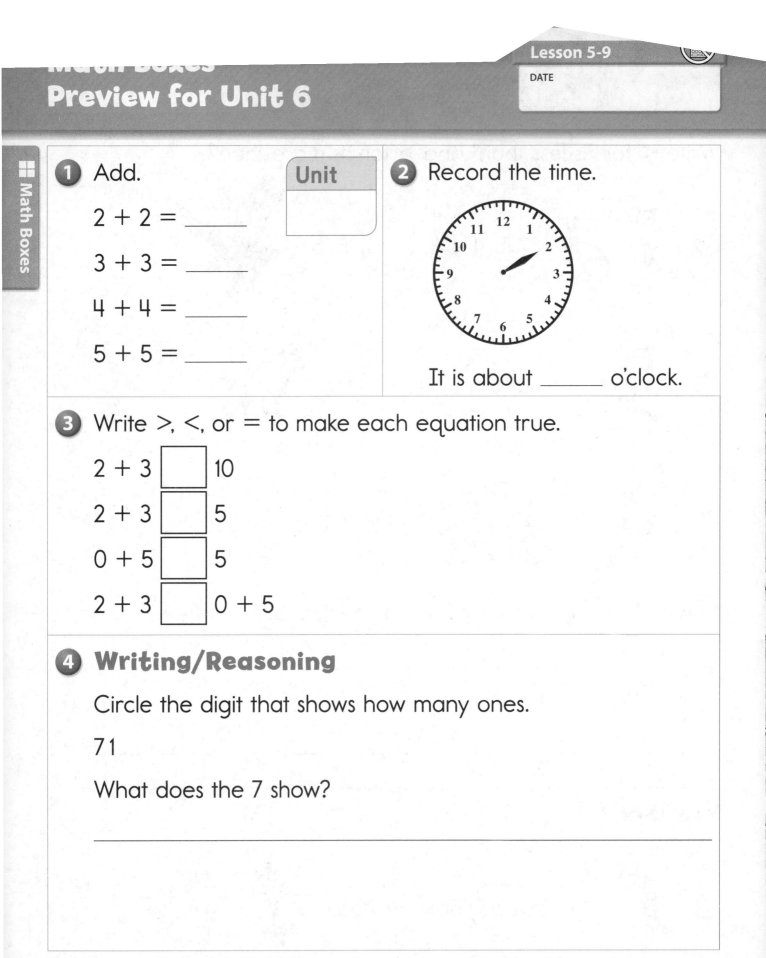

It is about _____ o'clock.

3 Write >, <, or = to make each equation true.

$2 + 3$ ☐ 10

$2 + 3$ ☐ 5

$0 + 5$ ☐ 5

$2 + 3$ ☐ $0 + 5$

4 **Writing/Reasoning**

Circle the digit that shows how many ones.

71

What does the 7 show?

_____

How Many More or Fewer?

Solve. Use the diagrams to help you.
Then write a number model to match.

1 John has 8 pennies. Nick has 2 pennies.

Who has more pennies? _____

How many more? _____

Number model: _____

Quantity

Quantity	
	_____
	Difference

2 June has 10 pennies. Mia has 6 pennies.

Who has fewer pennies? _____

How many fewer? _____

Number model: _____

Quantity

Quantity	
	_____
	Difference

3 Dante has 7 pennies. Kala has 15 pennies.

Who has more pennies? _____

How many more? _____

Number model: _____

Quantity

Quantity	
	_____
	Difference

4 For Problem 3, Jamal wrote $7 + \_\_\_\_\_ = 15$.
Explain how his number model matches the problem.

_____

_____

Making True Number Sentences

Write +, −, or = in the boxes to make true number sentences.

Example: 9 ☐ 5 ☐ 4

9 $-$ 5 $=$ 4

1 8 ☐ 7 ☐ 1

2 2 ☐ 8 ☐ 10

3 6 ☐ 5 ☐ 11

4 14 ☐ 9 ☐ 5

5 15 ☐ 6 ☐ 9

6 8 ☐ 12 ☐ 4

Try This

Some of the problems have more than one answer.

Example: 9 ☐ 5 ☐ 4

9 $-$ 5 $=$ 4 and 9 $=$ 5 $+$ 4

Circle the number sentences above that have more than one answer.

Write another number sentence for each problem you circled.

1 Callie has 3 pencils.
Lex has 9 pencils.
How many more pencils
does Lex have than Callie?

Choose the best answer.
- ⬭ 3 pencils
- ⬭ 5 pencils
- ⬭ 6 pencils
- ⬭ 7 pencils

2 Use your number grid
to add.

$25 + 20 = \boxed{}$

3 Look at your Pattern Block Template.

How many sides does a rectangle have?

_____ sides

How many sides do 2 rectangles have?

_____ sides

4 **Writing/Reasoning** Find the rule and fill in the frames.

Rule

3	6	9		

How did you find the rule?

Math Boxes

Math Boxes

1 ☐ = 8 + 0   3 + 0 = ☐

☐ + 4 = 4   0 + ☐ = 6

Unit

2 Use 1 paper clip.

Draw a pencil that is
3 paper clips long.

3 Use your number grid.

Start at 14.
Count back 8.

You end at _____.

14 − 8 = ☐

4 **Favorite Drinks of Angie's Friends**

How many of Angie's friends
like milk?

_____

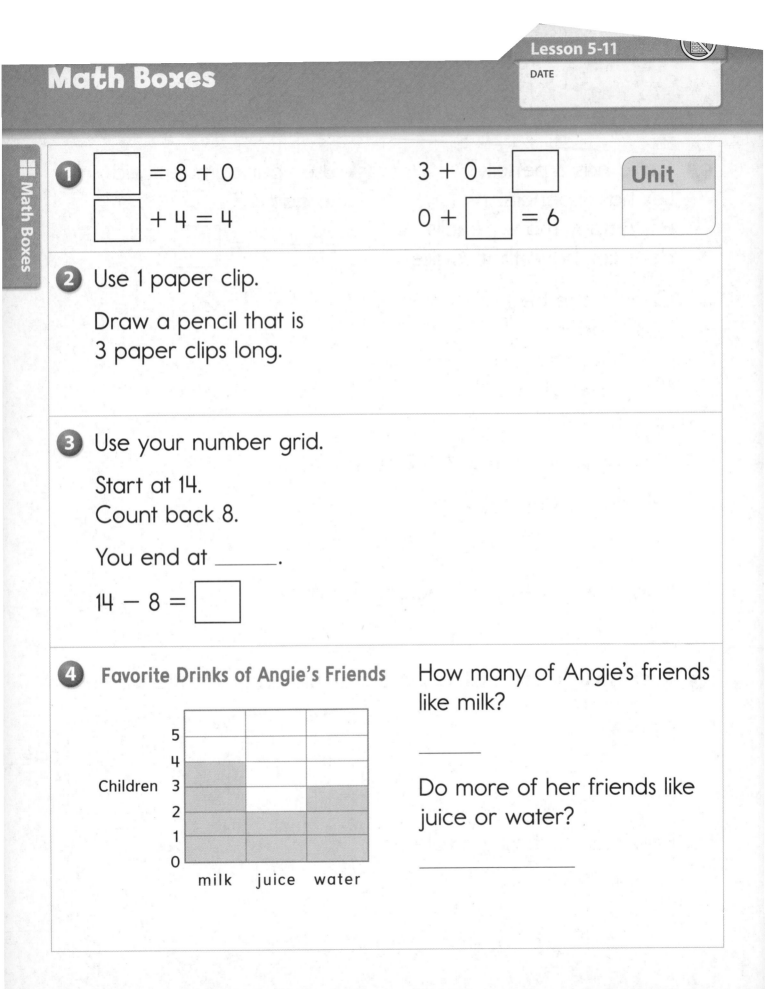

Children 5 4 3 2 1 0

milk   juice   water

Do more of her friends like
juice or water?

_____

Number Grid

Use the number grid.

Start at 44.
Count up 21 spaces.
Where do you land? _____

−9	−8	−7	−6	−5	−4	−3	−2	−1	0
1	2	3	4	5	6	7	8	9	10
11	12	13	14	15	16	17	18	19	20
21	22	23	24	25	26	27	28	29	30
31	32	33	34	35	36	37	38	39	40
41	42	43	44	45	46	47	48	49	50
51	52	53	54	55	56	57	58	59	60
61	62	63	64	65	66	67	68	69	70
71	72	73	74	75	76	77	78	79	80

Show your partner how you used the number grid.

Math Boxes

Math Boxes

1 Record the time.

A little after _____ o'clock

2 Use base-10 blocks to add.

10 + 14 = _____

Show your sum with | and ▪.

3 How many dots in all? _____ dots

How can you know without counting?

4 Fill in the rule and the missing numbers.

| Rule | 2 | 12 | 22 | | |

1 $\boxed{} = 3 + 7$

$6 + 4 = \boxed{}$

$\begin{array}{r} 2 \\ + 8 \\ \hline \boxed{} \end{array}$ $\begin{array}{r} 10 \\ + 0 \\ \hline \boxed{} \end{array}$

Unit

games

2 What is the value of the 0 in 10?

_____

What is the value of the 9 in 90?

_____

3 $12 + 13 = 25$

$10 + 15 = 25$

$12 + 13 \boxed{} 10 + 15$

Which symbol goes in the box?

Choose the best answer.

◯ <

◯ =

◯ >

4 Raj used his number line.

He started at 10.
He moved forward some hops.
He landed on 16.
How many hops did Raj make?

_____ hops

Math Boxes

Addition Fact	Know It	Don't Know It	How I Can Figure It Out . . .
0 + 1			
7 + 2			
0 + 3			
3 + 2			
0 + 5			
10 + 2			
3 + 1			
2 + 2			
6 + 0			
1 + 1			

Addition Fact	Know It	Don't Know It	How I Can Figure It Out . . .
4 + 1			
5 + 2			
4 + 0			
0 + 2			
1 + 5			
7 + 0			
10 + 1			
2 + 4			
1 + 6			
0 + 9			

Addition Fact	Know It	Don't Know It	How I Can Figure It Out . . .
1 + 2			
8 + 1			
0 + 0			
9 + 2			
2 + 6			
8 + 0			
7 + 1			
0 + 10			

Addition Fact	Know It	Don't Know It	How I Can Figure It Out . . .
10 + 10			
5 + 5			
10 + 5			
8 + 8			
7 + 10			
3 + 3			
8 + 2			
4 + 10			
6 + 4			
10 + 8			

Addition Fact	Know It	Don't Know It	How I Can Figure It Out . . .
7 + 7			
6 + 6			
3 + 10			
9 + 9			
1 + 9			
7 + 3			
6 + 10			
4 + 4			
10 + 9			

Notes

Notes

Number Cards 0-15

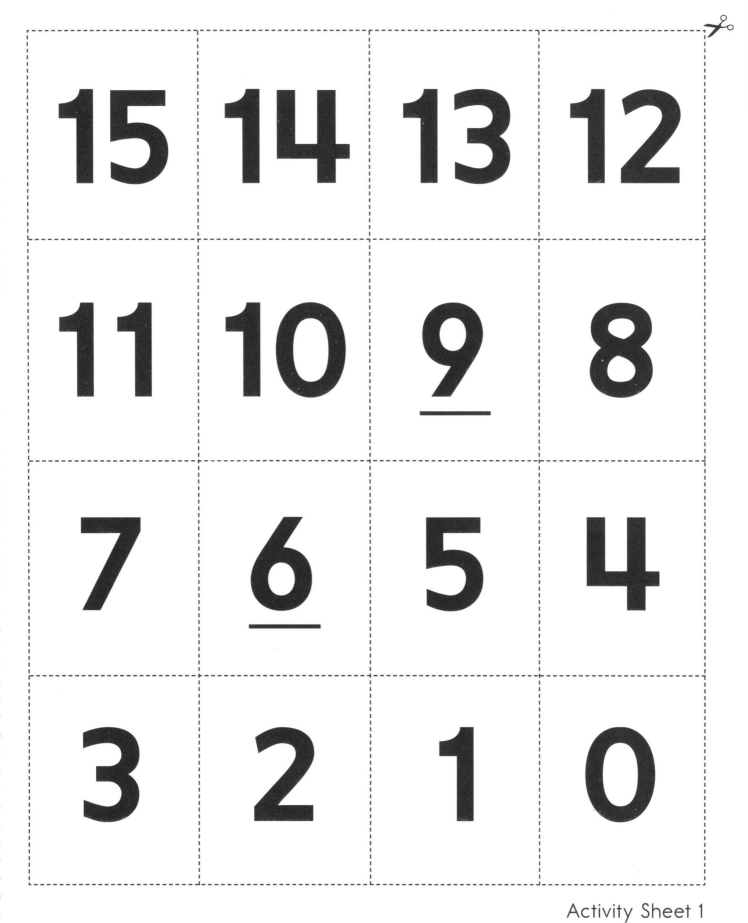

Number Cards 16–22 and Symbol Cards

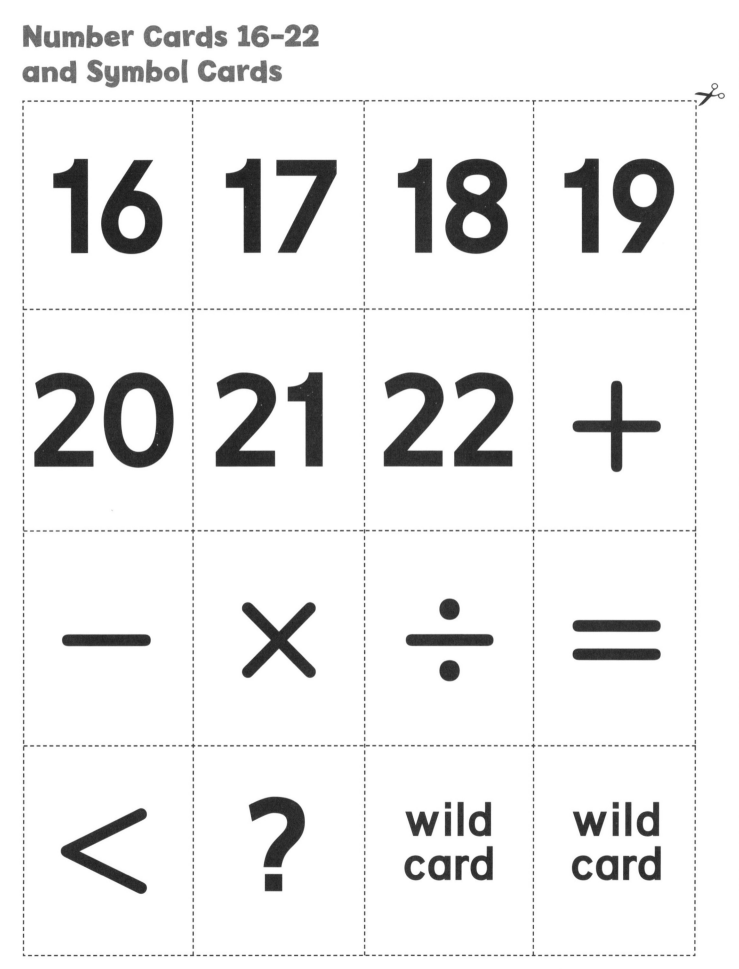

16	17	18	19
20	21	22	+
−	×	÷	=
<	?	wild card	wild card

Ten Frame

NAME   DATE

Activity Sheet 3

Double Ten Frame

Tens-and-Ones Mat

NAME   DATE

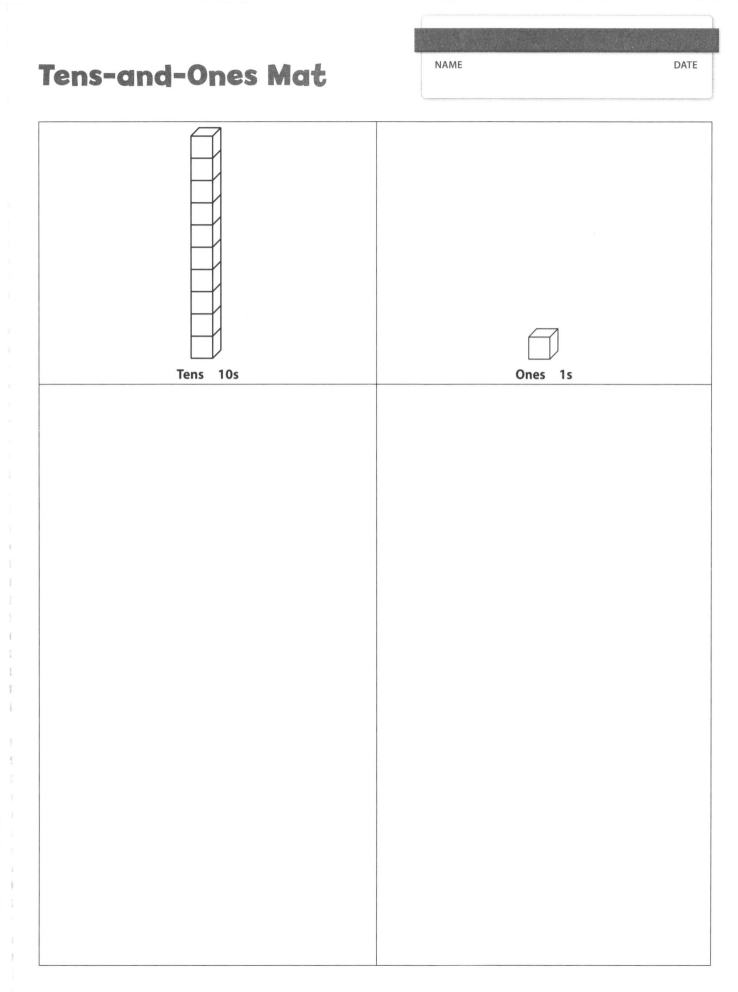

Tens   10s

Ones   1s

Animal Cards

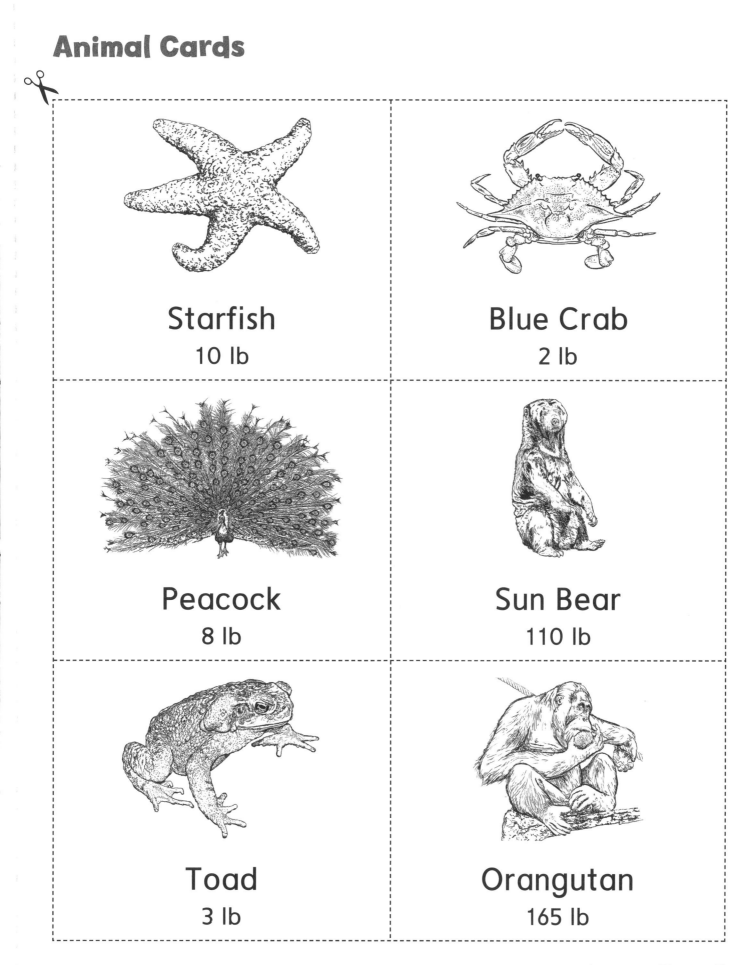

Starfish
10 lb

Blue Crab
2 lb

Peacock
8 lb

Sun Bear
110 lb

Toad
3 lb

Orangutan
165 lb

Animal Cards

Blue Crab
4 in.

Starfish
6 in.

Sun Bear
54 in.

Peacock
60 in.

Orangutan
48 in.

Toad
5 in.

Activity Sheet 5

Animal Cards

Parrot
3 lb

Squirrel
4 lb

Skunk
9 lb

Owl
5 lb

Flamingo
9 lb

Octopus
20 lb

Animal Cards

Squirrel
5 in.

Parrot
31 in.

Owl
20 in.

Skunk
8 in.

Octopus
36 in.

Flamingo
40 in.

Activity Sheet 6

Animal Cards

First-grade girl
41 lb

7-year-old boy
50 lb

Cheetah
120 lb

Porpoise
98 lb

Penguin
75 lb

Beaver
56 lb

Animal Cards

7-year-old boy
50 in.

First-grade girl
43 in.

Porpoise
72 in.

Cheetah
48 in.

Beaver
30 in.

Penguin
36 in.

Activity Sheet 7

Animal Cards

Cat
7 lb

Fox
14 lb

Koala
19 lb

Raccoon
23 lb

Rabbit
6 lb

Eagle
15 lb

Animal Cards

Fox
20 in.

Cat
12 in.

Raccoon
23 in.

Koala
24 in.

Eagle
35 in.

Rabbit
11 in.

Activity Sheet 8